A brief course in

semimicro qualitative analysis

third edition

A brief course in semimicro qualitative analysis

third edition

William E. Caldwell
Oregon State University

G. Brooks King
Washington State University

D. Van Nostrand Company
New York Cincinnati Toronto London Melbourne

D. VAN NOSTRAND Company Regional Offices:
Cincinnati New York

D. VAN NOSTRAND Company International Offices:
London Toronto Melbourne

ISBN: 0-442-21010-8

Published by D. VAN NOSTRAND Company

20 19 18 17 16 15 14 13

Preface to third edition

A number of changes have been made in this Third Edition to improve the usefulness of the text. Directions for separations and tests for both cations and anions have been carefully reviewed with some refinements made in directions to give more satisfactory results. The exercise and problem section has been completely revised with new and additional exercises. Answers to all numerical problems are given.

The authors gratefully acknowledge the help and suggestions of many users of the Second Edition for modifying experimental procedures. Special thanks are due Dr. A. B. Scott, of Oregon State University, and Dr. C. J. Nyman, of Washington State University, for their valuable assistance.

W. E. C.
G. B. K.

Contents

Part one Principles of qualitative analysis

Part one

Principles of qualitative analysis

An introduction to qualitative analysis

Qualitative analysis is that branch of analytical chemistry which determines the elements, radicals, or ions present in an unknown substance or mixture of substances. Quantitative analysis is concerned with the amount of an element or compound present. For example, a qualitative analysis may indicate the presence of lead, sulfur, silica, and arsenic; a quantitative analysis may show 20% lead, 25% sulfur, 2% arsenic, and 53% silica.

This manual will consider only *qualitative analysis*. One can sometimes distinguish qualitatively between two substances by some difference in the way they affect the senses—taste, sight, or smell. For example, salt and sugar differ in taste; carbon and sulfur differ in color and crystalline form. At times it is not so easy to detect differences, and then substances must be more closely examined for their chemical or physical properties. It is possible to differentiate the colorless salts $AgNO_3$ and $NaNO_3$ by such physical properties as melting point, crystalline form, or degree of solubility in water, for example. A more precise means of differentiation, however, is the addition of HCl to their water solutions; a white, curdy precipitate (AgCl) is formed with one, and no precipitate with the other.

There is greater difficulty in identifying the various constituents of a mixture. The comparative solubility of different substances is the principal basis for their separation. The difference might be in *physical solubility*, such as NaCl in H_2O or I_2 in CCl_4, or in *chemical solubility*, such as $CaCO_3$ in HCl or AgCl in NH_4OH.

The usual procedure for testing an unknown is dissolving the sample and testing the resulting solution for the ions which may be present. An alloy, for example, when acid dissolved gives a solution of the metals as ions (cations). Dissolving of an acid, base, or salt gives respective cations and anions of the elements or radicals present. Thus, qualitative analysis is commonly divided into *anion analysis* and *cation analysis*. Test procedures involve a separation of the ions by precipitation and solution, followed by specific tests for each ion. The specific tests may give characteristically colored solutions or precipitates on the addition of certain reagents. For example, the addition of $K_4Fe(CN)_6$ (potassium ferrocyanide) to a solution containing Fe^{+++} gives a deep blue precipitate of $KFe[Fe(CN)_6]$ called Prussian blue.

An analysis performed on a solution of the unknown sample may not always indicate how the ions were originally combined. Obviously, if only Na^+ and Cl^- are found, the original substance must have been NaCl. If Na^+, K^+, Cl^-, and NO_3^- are found, the original may have been a mixture of NaCl and KNO_3, or of $NaNO_3$ and KCl, or all four salts. To identify the substances present in such a mixture, additional tests must be made; these are usually of a physical nature, such as examination of crystal structure, color, and so on.

Processes for the identification of minerals can be considered as a part of qualitative analysis, and they played a part in the historic development of analytical methods. Methods were physical or chemical in scope. One might consider a small air blowpipe as a physical instrument to aid in attaining a higher-temperature jet flame to bring about an oxidative chemical change. For example, mineralogical chemists can use a charcoal block with a small indentation into which is placed a fragment of a mineral. The mineral is then heated by a small jet of flame augmented in intensity by mouth use of an air blowpipe. Such a mineral as As_2S_3 is oxidized to white fumes of As_2O_3, which has a somewhat garlic-like odor; zinc sulfide is oxidized to white ZnO which when heated with a drop of $Co(NO_3)_2$ gives a residue of green $CoO \cdot ZnO$.

The blowpipe can be considered a very simple instrument compared to modern instruments and methods for the quite rapid qualitative and quantitative analysis of diverse unknowns. The unknown may be mineral, alloy, acid, base, salt, or organic material. Instrumental methods are briefly discussed on page 165. Schemes of chemical separations and testings of cations and anions in this manual supply subject-matter background for the possible instrumental analysis of an advanced laboratory or industry.

Many industrial-process separations are based on principles portrayed in qualitative analysis. One specific example is the purification of aluminum ore, which may be considered a mixture of aluminum hydroxide and iron hydroxide. NaOH will dissolve $Al(OH)_3$ and not $Fe(OH)_3$ (see similar chemical treatment in an analytical scheme to follow). Basic chemistry of both nonmetals and metals is encountered in analytical chemical changes and procedures.

General principles

Many of the general principles and concepts learned in elementary chemistry are applicable to qualitative analysis. Since such principles are covered in general chemistry texts, this manual will review briefly only those topics which are particularly pertinent to qualitative procedures. The more important topics to be considered are:

- Manner of expressing amounts of a solute
- Methods of expressing the concentration of a solution
- Law of mass action and chemical equilibrium
- Ionic nature of acids, bases, and salts in water solution
- Exponential numbers
- Ionic equilibrium
  - Ionization constants
  - Common ion effect
  - Buffer action
  - Hydrogen ion concentration, *p*H
  - The ionization of water
- Logarithmic calculations of *p*H and *p*OH
- Solubility product principle
  - The formation of precipitates
  - Dissolving of precipitates
  - Precipitation of cations as sulfides
- Valence and oxidation number
- Chemical changes in qualitative analysis
  - Ionic union
  - Oxidation-reduction
  - Oxidation potentials
  - Formation of double salts
  - Formation of complex salts and complex ions
  - Amphoterism
  - Hydrolysis
  - Hydrolysis constants

1
Manner of expressing amounts of a solute

Gram moles (moles). Amounts of a solute may be expressed according to units of weight, such as so many grams or pounds. It is more common in making solutions of chemicals to take into consideration gram molecular weights of the solute. A gram molecular weight is that number of grams numerically equal to the molecular or formula weight. An abbreviation for the term *gram molecular weight* is *gram mole* or simply *mole*. In this manual the term *mole* will be used in place of *gram molecular weight*. Sometimes in industry *pound mole* is used, meaning *pound molecular weight*. As well as referring to a mole of a substance, one may refer to a number of moles, such as 7 moles, 0.3 mole, 0.126 mole, and so on.

Some problems necessitate change of grams to moles or moles to grams:

Problem 1. A 6.8 g amount of $CaSO_4$ is how many moles of $CaSO_4$?

$$136 \text{ g of } CaSO_4 = 1 \text{ mole}$$

$$6.8/136 = 0.05 \text{ mole of } CaSO_4$$

Problem 2. How many grams of H_3PO_4 are represented by 3.75 moles H_3PO_4?

$$98 \text{ g of } H_3PO_4 = 1 \text{ mole}$$

$$98 \times 3.75 = 367.5 \text{ g } H_3PO_4$$

Gram ion. Another important term is *gram ion*, which is a manner of expressing weight of ionic quantities. That number of grams numerically equal to the sum of the weights of the atoms in an ion (ionic weight) is a gram ion. When 159.6 g of $CuSO_4$ (1 mole) is present in dilute solution, it may be assumed to contain 63.6 g or 1 g ion of Cu^{++} ion, and 96 g or 1 g ion of SO_4^{--} ion. Similarly 4 moles of $Al_2(SO_4)_3$ contain 8 g ions of Al^{+++} and 12 g ions SO_4^{--}. Fractional parts of gram ions may, of course, be referred to.

Problem 3. A 0.5 g quantity of $Cu(NO_3)_2$ in water solution contains how many gram ions of Cu^{++} and how many of NO_3^-?

187.6 g of $Cu(NO_3)_2$ = 1 mole and contains 1 g ion Cu^{++} and 2 g ions NO_3^-

$$0.5 \text{ g of } Cu(NO_3)_2 = \frac{0.5}{187.6} = 0.00266 \text{ mole}$$

0.00266 mole of $Cu(NO_3)_2$ contains 0.00266 g ion Cu^{++} and 0.00532 g ion NO_3^-.

2
Methods of expressing the concentration of a solution

Percentage composition. The weight of a solute in a given weight of solution may be expressed in percentage—that is, parts of solute per 100 parts of solution. Thus a solution containing 5 g of solute and 95 g of solvent would be a 5% solution, a 25% solution would be a ratio of 25 parts of solute to 75 parts of solvent, and so on.

Molarity. The molarity of a solution is defined as the number of moles of solute per liter of solution. A solution containing 1 mole of solute per liter of solution is 1 molar (1 *M*), one containing 0.5 mole per liter is 0.5 *M*, and so on. Molar solutions are usually made by dissolving the required amount of solute in a small volume of water and then diluting in a volumetric container up to a liter in total volume. Hence the weight of solute is known, the total volume is known, but neither the volume nor weight of solvent is known.

The following formulations are useful in calculations involving molarity:

$$\text{molarity} = \frac{\text{moles of solute}}{\text{liters of solution}}$$

Since

$$\text{moles} = \frac{\text{grams of solute}}{\text{mol. wt. of solute}}$$

then

$$\text{molarity} = \frac{\text{grams}}{\text{mol. wt.} \times \text{liters of solution}} \qquad (1)$$

Study the following illustrative problems:

Problem 1. Calculate the molarity of a solution containing 10 g of H_2SO_4 in 500 ml of solution.

10 g in 500 ml is equivalent to 20 g in 1 liter.

20 g of H_2SO_4 (mol. wt. 98) is 20/98 or 0.204 mole.

Therefore, the molarity is 0.204.

The molarity may also be calculated directly from (1) above:

$$\text{molarity} = \frac{10 \text{ (grams)}}{98 \text{ (mol. wt.)} \times 0.5 \text{ (liters)}} = 0.204$$

Problem 2. Calculate the weight of H_2SO_4 (in grams) in 2 liters of 0.1 *M* solution.

One liter of 0.1 *M* H_2SO_4 contains 0.1×98 or 9.8 g H_2SO_4. Two liters then would contain 2×9.8 or 19.6 g. Or, by substituting in (1) above

$$0.1 \text{ (molarity)} = \frac{x \text{ (grams)}}{98 \text{ (mol. wt.)} \times 2 \text{ (liters)}}$$

Then

$$x = (0.1)(98)(2) = 19.6 \text{ grams}$$

Normality. Normality is defined as the number of equivalents per liter of solution. A *normal solution* is made in the same manipulative way as a molar solution except that a gram equivalent weight, rather than a gram molecular weight, of solute is dissolved and diluted with solvent to a total volume of 1 liter. The gram equivalent weight of an acid contains 1 gram atom of replaceable hydrogen (1.008 g); the gram equivalent weight of a base or salt contains the hydrogen equivalent of the metal.* The hydrogen equivalent of a metal is the gram atomic weight of the metal divided by its valence.

Examples. The equivalent weight of sodium in NaCl (atomic weight of Na is 23 and valence is 1) is 23; that of copper in $CuCl_2$ (atomic weight of Cu is 63.6, valence 2) is 31.8; that of aluminum in $AlCl_3$ (atomic weight of Al is 27, valence 3) is 9.0. It is apparent that 1 gram atom of sodium is equivalent to 1 gram atom of hydrogen, while 1 gram atom of copper is equivalent to 2 gram atoms of hydrogen, and 1 gram atom of aluminum is equivalent to 3 of hydrogen.

The compositions of several 1 *N* solutions are given below as illustrations:

One liter of 1 *N* HCl contains 36.5 g of HCl, which in turn contains 1 gram atom of replaceable hydrogen.

One liter of 1 *N* H_3PO_4 contains 32.7 g of H_3PO_4. The molecular weight is 98; since the mole furnishes 3 gram atoms of hydrogen, the gram equivalent weight is 98/3 or 32.7 g.

One liter of 1 *N* $CuSO_4$ (mol. wt. 159.6) contains 79.8 g of the salt since that amount contains 31.8 g (the hydrogen equivalent) of copper.

* This method of computing equivalent weights is not applicable to substances to be used in oxidation-reduction reactions. The equivalent weight of an oxidizing or reducing agent is the molecular or ionic weight divided by the number of electrons transferred per mole of substance. Thus in the reaction

$$5\,H_2C_2O_4 + 2\,KMnO_4 + 3\,H_2SO_4 \rightarrow 2\,MnSO_4 + K_2SO_4 + 10\,CO_2 + 8\,H_2O$$

5 electrons are transferred per mole of $KMnO_4$ and 2 electrons per mole of $H_2C_2O_4$; hence the equivalent weight of $KMnO_4$ = mol. wt./5 and that of $H_2C_2O_4$ is mol. wt./2.

A formulation for making normality calculations similar to the one above for molarity may be derived:

$$\text{normality} = \frac{\text{equivalents of solute}}{\text{liters of solution}}$$

Since $$\text{equivalents} = \frac{\text{grams of solute}}{\text{equiv. wt. of solute}}$$

then $$\text{normality} = \frac{\text{grams}}{\text{equiv. wt.} \times \text{liters of solution}} \qquad (2)$$

Study the following examples:

Problem 3. What is the normality of a solution which contains 4.9 g of H_2SO_4 in 250 ml of solution?

Molecular weight of $H_2SO_4 = 98$. Equivalent weight $= 98/2 = 49$. 4.9 g $=$ 0.1 equivalent.

0.1 equivalents in 250 ml (1/4 liter) would be the same as 0.4 equivalent in 1000 ml (1 liter). Therefore the normality is 0.4. Or, substituting in (2) above, we get

$$\text{normality} = \frac{\text{grams of solute}}{\text{equiv. wt. of solute} \times \text{liters}} = \frac{4.9}{49 \times 0.25} = 0.4$$

Problem 4. What weight of NaOH is present in 5 liters of 0.2 N solution?

Molecular weight of NaOH $= 40$. Equivalent weight $= 40/1 = 40$.

One liter of 0.2 N solution would contain $40 \times 0.2 = 8$ g of NaOH; 5 liters would contain $5 \times 8 = 40$ g of NaOH. Or, substituting in (2) above,

$$N = \frac{\text{grams}}{\text{equiv. wt.} \times \text{liters}}$$

$$0.2 = \frac{\text{g}}{40 \times 5}$$

$$\text{grams of NaOH} = (0.2)(40)(5) = 40$$

Note that in this problem the molarity of the solution is the same (0.2) as the normality since the equivalent weight of NaOH is the same as its molecular weight. Since the molecular or formula weight of a compound always bears a small whole-number relationship (valence) to the equivalent weight of the compound, the normality of a solution will always either be equal to the molarity or a simple multiple of it—that is, 1, 2, 3, and so on.

Another useful relationship in normality calculations is the following:

$$N_1 \times V_1 = N_2 \times V_2 \qquad (3)$$

N_1 and V_1 represent the normality and volume, respectively, of one solution, and N_2 and V_2 represent the normality and volume of a second solution. Study the following sample calculations:

Problem 5. Determine the volume of 16 *N* nitric acid to be diluted to make 250 ml of 2 *N* nitric acid solution.

Substituting in (3) above:

$$16 \times V_1 = 2 \times 250 \text{ ml}$$

Then

$$V_1 = \frac{2 \times 250 \text{ ml}}{16} = 31.25 \text{ ml}$$

Problem 6. Calculate the normality of an acid 10 ml of which exactly neutralizes 30 ml of a base, 0.125 *N*.

$$N_1 \times 10 = 0.125 \times 30$$

$$N_1 = \frac{0.125 \times 30}{10} = 0.375$$

Molality. The molality of a solution is defined as the number of moles of solute per 1000 g of solvent. The composition of a *molal solution* is known with regard to weight of both solute and solvent and is thus independent of temperature changes. This means of expressing the concentration of a solution is used in the study of properties of solutions in which the ratio of solute particles to solvent particles is a factor, for example, in computations involving the lowering of the freezing point or the rise in boiling point of solutions. Molal and molar solutions are nearly identical if the solutions are dilute, but in more concentrated solutions wide differences may be expected.

Test solutions. Test solutions containing metal ions for semimicro qualitative analysis are made to contain 10 mg of metal ion per milliliter. This would be 10 g of metal ion per liter. For example, in making up a Cu^{++} ion test solution using $Cu(NO_3)_2$, the gram molecular weight of which is 187.6 g and which contains 63.6 g of Cu^{++}, one would use $187.6/63.6 \times 10 = 29.5$ g or the weight of $Cu(NO_3)_2$ which would contain 10 g of Cu^{++}; 29.5 g of $Cu(NO_3)_2$ dissolved in a liter of solution would contain 10 mg Cu^{++} per milliliter.

Example. When preparing a Sn^{++} ion test solution from $SnCl_2 \cdot 2\,H_2O$, what weight of the salt should be used per liter of solution?

One mole of $SnCl_2 \cdot 2\,H_2O$ weighs 225.7 g and contains 118.7 g of Sn^{++} ion. To obtain 10 g of Sn^{++} ion per liter of solution, one would use $10/118.7 \times$

225.7 = 19.01 g of the salt. Salts, such as $SnCl_2 \cdot 2\,H_2O$, which are subject to hydrolysis require acid to aid in their solution. Consequently, in making up a test solution of Sn^{++}, 19 g of $SnCl_2 \cdot 2\,H_2O$ is dissolved in enough 6 *M* HCl to give 1 liter of solution. This solution then contains 10 g of Sn^{++} per liter, or 10 mg Sn^{++} per ml.

Solutions of slightly soluble substances. Qualitative analytical separations are best made when one obtains one ion in a very soluble state and the other as part of a very slightly soluble substance. The solubility of slightly soluble substances may be expressed in grams per liter or in molarity. The expression of the concentration of very slightly soluble substances offers a problem in the use of exponentials. (For a review of exponential arithmetic, see page 17.)

Problem 7. It has been determined experimentally that 1 liter of pure water will dissolve 0.0025 g of $BaSO_4$. It could also be stated that 1 liter will dissolve 2.5×10^{-3} g $BaSO_4$. The concentration of $BaSO_4$ in a saturated solution may also be expressed as molarity.

Molecular weight of $BaSO_4 = 233$.

$$M = \frac{0.0025 \text{ g/liter}}{233 \text{ g/mole}} = 0.0000107 \text{ moles/liter}$$

or, expressed exponentially,

$$M = 1.07 \times 10^{-5} \text{ moles/liter}$$

Problem 8. Ag_3PO_4 is soluble in the small amount of 0.0065 g/liter. What is the molarity of a saturated silver phosphate solution?

One mole of $Ag_3PO_4 = 418.4$ g.

$$M = \frac{0.0065 \text{ g/liter}}{418.4 \text{ g/mole}} = 0.0000155 = 1.55 \times 10^{-5} \text{ moles/liter}$$

3
Law of mass action and chemical equilibrium

Factors influencing speed of reaction. The speed at which substances interact chemically depends on a number of factors:

1. *Nature of reacting substances.* Metals and nonmetals may be arranged in an activity series which is an index of their combining or replacement tendencies. For example, magnesium acts more rapidly with dilute acid than iron; chlorine replaces bromine from a bromide.

2. *Catalysts.* The speed of a chemical change may be altered by a catalyst.

3. *Temperature.* In general, an increase in temperature of 10° C approximately doubles the speed of a chemical change.

4. *Surface factors and solution.* Increased surface area of a solid or a liquid undergoing union with a gas increases the rate of change. A given quantity of sulfur spread over an area with a large, exposed surface will burn faster than if a cube of the same size were burning (other conditions being equal). Two mixed dry solids which undergo a very slow chemical change may react very rapidly when a mutual solvent is added to dissolve them.

5. *Concentration.* Kinetically, chemical changes take place as a result of molecular or ionic collisions. The greater the number of collisions per unit of time, the greater the rate of chemical change. Therefore, gases under pressure (molecules closer together, concentration higher) undergo relatively high rates of change. This also may be said for concentrated solutions. It should be understood, however, that collisions do not necessarily result in chemical change. Temperature and the nature of the substances have a bearing on the effectiveness of collisions; in some cases a high percentage of collisions, in other cases a low percentage of collisions, result in chemical change. Actual contact between interacting substances is, of course, essential for chemical reaction.

The law of mass action and equilibrium constants. After compiling data on the relationship between concentration and rate of chemical change, Guldberg and Waage in 1867 enunciated the law of mass action:

> *The speed of a chemical change is proportional to the concentrations (or active masses) of the reacting units (ions, atoms, or molecules).*

From the law of mass action, the equilibrium constant for an equilibrium reaction may be derived. For a development of the law of mass action and its application to various chemical changes, the student is referred to a college-level general chemistry text.

For the general equilibrium reaction

$$mA + nB \leftrightarrows rC + sD$$

in which m, n, r, and s represent the coefficients of A, B, C, and D, respectively, in the balanced equation

$$K = \frac{[C]^r[D]^s}{[A]^m[B]^n}$$

where K is the *equilibrium constant.* In other words, *at equilibrium, the product of the concentrations of products divided by the product of concentrations of reactants—each concentration raised to that power which corresponds to the coefficient in the balanced chemical equation—is a constant.*

Examples.

$$H_2 + I_2 \leftrightarrows 2\ HI \qquad K = \frac{[HI]^2}{[H_2][I_2]}$$

$$N_2 + 3\ H_2 \leftrightarrows 2\ NH_3 \qquad K = \frac{[NH_3]^2}{[N_2][H_2]^3}$$

$$4\ NH_3 + 5\ O_2 \leftrightarrows 4\ NO + 6\ H_2O \qquad K = \frac{[NO]^4[H_2O]^6}{[NH_3]^4[O_2]^5}$$

Reactions which go to completion.* Reactions which proceed in only one direction are complete, while those in which an equilibrium exists are incomplete. Every reaction becomes complete if one of the products of the reaction is removed as fast as it is formed. There are three conditions under which a reaction may proceed to practical completion.

1. FORMATION OF A PRECIPITATE. Assume that two salts are dissolved separately in water and that the two solutions are mixed. Since both the reactants and the products can be considered to be completely ionized, no reaction occurs unless one of the product salts is quite insoluble in water. A case of this kind would be the addition of a solution of silver nitrate to one of sodium chloride.

$$AgNO_3 + NaCl \rightarrow AgCl\downarrow + NaNO_3$$

or

$$Ag^+ + Cl^- \rightarrow AgCl\downarrow$$

It is not correct to say that the above chemical change is entirely complete to the

* Throughout this book a single arrow is used to indicate a complete or practically complete chemical change. Double arrows denote an equilibrium reaction. If the double arrows are of unequal length, the longer arrow indicates whether products or reactants are in greater proportion at equilibrium.

right, because AgCl is soluble in water to a very slight extent. In another solvent, such as liquid ammonia, AgCl is quite soluble, and the reaction does not go to completion even in the approximate sense. Since in qualitative analysis we usually deal with water solutions, the importance of solubility rules in water in regard to completeness of many chemical changes at once becomes apparent.

2. FORMATION OF A GAS. When an insoluble or a very slightly soluble gas is a product of a chemical change, the change may be considered complete. For example,

$$Na_2S + 2\ HCl \rightarrow H_2S\uparrow + 2\ NaCl$$

or

$$S^{--} + 2\ H^+ \rightarrow H_2S\uparrow$$

3. FORMATION OF A SLIGHTLY IONIZED PRODUCT. Chemical changes in which water or some other slightly ionized product is formed are classed as practically complete chemical changes.

$$H^+ + OH^- \rightarrow H_2O$$

$$H^+ + C_2H_3O_2^- \rightarrow HC_2H_3O_2$$

The law of mass action applies to reactions in which a precipitate, gas, or slightly ionized product is formed in the following way. The precipitate, gas, or slightly ionized product is removing ions or molecules from the zone where collisions are taking place. The concentration of these ions or molecules is greatly reduced in the reacting zone. Since the speed of the reverse action is proportional to the concentrations of products, and since the concentration of one product is almost zero because of precipitation, gas formation, or slightly ionized product, the speed of reverse chemical change is almost zero.

4
Ionic nature of acids, bases, and salts in water solution

According to the modern electron theory of valency, atomic union in compounds is explained by electrovalency, covalency, or coordinate valency.

Salts. The best examples of electrovalency are salts. Metal atoms give a number of electrons equivalent to their valence to nonmetal atoms in chemical union to form salts. The solid salts are usually composed of positive metallic and negative nonmetallic ions in a crystal lattice. When a crystal of a salt dissolves, the respective ions break away from the solid and are dispersed in the solvent. To differentiate between a solid salt and its solution, we shall represent a solid salt by a molecular formula and a solution of the salt by its ions. For example,

$$\underset{\text{solid}}{CuCl_2} \rightleftharpoons \underset{\text{in solution}}{Cu^{++} + 2\ Cl^-}$$

Acids. Acids are covalent compounds, which means that there is a sharing of an electron pair between a hydrogen atom and the negative part of the acid; the hydrogen atom and the negative part each contribute an electron to this pair. A pure acid is composed, then, not of ions but of molecules. It forms ions only by reaction with a solvent:

$$HCl + H_2O \rightleftharpoons H_3O^+ + Cl^-$$

$$HCl + NH_3 \rightleftharpoons NH_4^+ + Cl^-$$

H_3O^+ is called the *hydronium ion* and may be written as $H^+ \cdot H_2O$; it can also be called a *hydrated hydrogen ion* or *hydrated proton.* A modern definition of an acid is that it is a *proton donor.* For simplification, water of hydration may be omitted in the writing of equations. For example,

$$HCl \rightleftharpoons H^+ + Cl^-$$

The extent to which various acids act with a solvent to form ions varies markedly with the acid. It is usual to call those acids which appear to ionize highly, strong acids; HCl, H_2SO_4, and HNO_3 apparently ionize over 90% in 0.1 M solution. On the other hand, acetic acid is only about 1.34% ionized at 20°C in 0.1 M solution and is therefore classified as a weak acid. Other weak acids are H_2S, HCN, and HF.

Bases. Most metal hydroxides are very slightly soluble in water, and thus do not yield a high hydroxide ion concentration, or basicity, in aqueous solutions. The water-soluble hydroxides NaOH, KOH, and LiOH are ionized even in the solid state. Since aqueous solutions of the latter hydroxides give a high concentration of hydroxide ions, they are classified as strong bases. Ammonium hydroxide (NH_4OH) gives relatively few hydroxide ions and therefore is classed as a weak base.

5

Exponential numbers

The use of exponentials. In chemistry, very large and very small numbers are usually expressed exponentially. To do this, a number is expressed as the product of two numbers, one of which is a multiple or "power" of 10, thus:

$$10 = 1 \times 10^1$$

$$100 = 1 \times 10^2$$

$$500 = 5 \times 100 = 5 \times 10^2$$

$$20{,}000 = 2 \times 10{,}000 = 2 \times 10^4$$

The first number in the product is termed the *coefficient*, and the second is a *power* of 10; for example, in the expression 5×10^2, 5 is the coefficient and 2 (written as a superscript) is the power or *exponent*. The exponent represents the number of decimal places which must be marked off to the *left* so that the first digit only is followed by the decimal point. For example, when we write the number 500, the decimal point is actually at the far right—that is, 500.—although we may not put it in. It is conventional in expressing a number exponentially to place the decimal point so that the coefficient is from 1 to 10. 500 would be shown as 5×10^2; the exponent 2 gives us the number of places (2) which we must mark off to the left, so that the coefficient in the expression is 5., which is between 1 and 10.

Examples.

$$500. = 5 \times 10^2$$

↑__┘

$$2{,}000{,}000. = 2 \times 10^6$$

↑__⑥__┘

$$517000. = 5.17 \times 10^5$$

↑__⑤__┘

Of course, 517000 might be expressed correctly as $517. \times 10^3$, or 51.7×10^4, or 0.517×10^6, and so on, but it is conventional to express this number as 5.17×10^5 so that, as we noted above, the coefficient will be between 1 and 10.

Similarly, small numbers may be expressed in exponential form, thus:

$$0.1 = \frac{1}{10} = 1 \times 10^{-1}$$

$$0.001 = \frac{1}{1000} = \frac{1}{10^3} = 1 \times 10^{-3}$$

$$0.00045 = 4.5 \times 10^{-4}$$

In these cases the *exponent* gives the number of decimal places which we must mark off to the *right* to give a coefficient between 1 and 10. Thus:

$$.00045 = 4.5 \times 10^{-4} \qquad .000000372 = 3.72 \times 10^{-7}$$

(4) (7)

Four types of operations with exponential expressions are commonly carried out in dealing with chemical problems:

Multiplication. In multiplying exponential expressions, multiply the coefficients in the usual manner and *add the exponents*, thus:

$$(4 \times 10^2) \times (2 \times 10^6) = 8 \times 10^8$$

$$(1.5 \times 10^3) \times (4 \times 10^2) = 6.0 \times 10^5$$

$$(1 \times 10^{-2})(3 \times 10^5) = 3 \times 10^3$$

$$(6 \times 10^4)(3 \times 10^{-1}) = 18. \times 10^3 = 1.8 \times 10^4$$

$$(5 \times 10^{-8})(4 \times 10^{-4}) = 20. \times 10^{-12} = 2.0 \times 10^{-11}$$

Division. In dividing exponential expressions, divide the coefficients and *subtract the exponents*; that is, subtract the exponent in the denominator from the one in the numerator, thus:

$$\frac{4 \times 10^4}{2 \times 10^3} = 2 \times 10^1$$

$$\frac{6.5 \times 10^8}{2} = \frac{6.5 \times 10^8}{2 \times 10^0} = 3.25 \times 10^8$$

$$\frac{3 \times 10^4}{1.5 \times 10^{-2}} = 2 \times 10^6$$

$$\frac{6 \times 10^{-6}}{2 \times 10^{-3}} = 3 \times 10^{-3}$$

$$\frac{8.6 \times 10^{-5}}{2.15 \times 10^3} = 4.0 \times 10^{-8}$$

$$\frac{1 \times 10^{-6}}{4 \times 10^3} = 0.25 \times 10^{-9} \quad \text{or} \quad 2.5 \times 10^{-10}$$

Extracting square and cube roots. Treat the coefficient in the usual way and *divide the exponent* by the root:

$$\sqrt{1 \times 10^6} = 1 \times 10^3$$

$$\sqrt[3]{8 \times 10^{-12}} = 2 \times 10^{-4}$$

$$\sqrt{16 \times 10^{-8}} = 4 \times 10^{-4}$$

$$\sqrt[4]{1 \times 10^{-8}} = 1 \times 10^{-2}$$

The exponential should be in such a form that the exponent gives an integral number when divided by the root, for example:

$$\sqrt{1.6 \times 10^{7}} \quad \text{change to} \quad \sqrt{16 \times 10^{6}}$$

or

$$\sqrt[3]{6.4 \times 10^{-8}} \quad \text{change to} \quad \sqrt[3]{64 \times 10^{-9}}$$

Raising an exponential form to some power. Raise the coefficient to the proper power and *multiply the exponents*, thus:

$$(1 \times 10^{4})^{2} = 1 \times 10^{8}$$

$$(6 \times 10^{-3})^{3} = 6^{3} \times (10^{-3})^{3} = 216. \times 10^{-9} = 2.16 \times 10^{-7}$$

$$(2 \times 10^{4})^{4} = 16. \times 10^{16} = 1.6 \times 10^{17}$$

See page 173 for problems and exercises dealing with manipulations of exponential expressions described above.

6
Ionic equilibrium

The Law of Mass Action may be applied to ionic as well as to molecular equilibria. The former would include the equilibrium between an acid and the ions which are formed when the acid reacts with the solvent, and also the equilibrium between a solid salt and the ions of the salt in aqueous solution (saturated). In this section we shall be concerned primarily with the equilibria in aqueous solutions of weak acids and bases.

Ionization constants of acids. Just as an equilbrium constant K may be established for molecular reactions, an *ionization constant* K_i may be determined for the equilibrium between the molecules of a weak electrolyte and its ions. Ionization constants for weak acids and bases may be designated as K_a and K_b, respectively. Strong electrolytes (salts, acids such as HCl, H_2SO_4, and HNO_3, and bases such as NaOH and KOH) are so nearly completely ionized that equilibrium calculations can no longer be applied.

Study the following calculations:

It has been determined experimentally that 0.1 M $HC_2H_3O_2$ at 20° C is only 1.34% ionized.

$$HC_2H_3O_2 + H_2O \rightleftharpoons H^+{\cdot}H_2O + C_2H_3O_2^-$$

or simply

$$HC_2H_3O_2 \rightleftharpoons H^+ + C_2H_3O_2^-$$

$$K_i \text{ or } K_a = \frac{[H^+] \times [C_2H_3O_2^-]}{[HC_2H_3O_2]}$$

If 0.1 M $HC_2H_3O_2$ at 20° C is 1.34% ionized,

$$[H^+] = 0.1\ M \times 1.34\% = 0.1 \times 0.0134 = 0.00134 \text{ g ion/liter}$$

$$[C_2H_3O_2^-] = 0.1\ M \times 1.34\% = 0.1 \times 0.0134 = 0.00134 \text{ g ion/liter}$$

$$[HC_2H_3O_2] = 0.1\ M \times 98.66\% = 0.1 \times 0.9866 = 0.09866 \text{ g mole/liter}$$

$$K_a = \frac{0.00134 \times 0.00134}{0.09866} = 0.0000182 = 1.82 \times 10^{-5}$$

The low value of this ionization constant for acetic acid is indicative of the fact that at equilibrium, when the speeds of ionization and of ionic union are equal, the extent of reaction in the forward direction is small. Acetic acid typifies a weak acid.

A table giving the ionization constants of a number of weak acids is given in the Appendix. It may be noted that two constants are given for carbonic acid and three for phosphoric acid. It will be recalled that acids with more than one replaceable hydrogen ionize in steps. Each successive step of the ionization is an equilibrium and thus would have a value for K. K_1 is for the first ionization step, K_2 for the second, and so on. In the case of H_3PO_4:

$$H_3PO_4 \rightleftharpoons H^+ + H_2PO_4^- \qquad K_1 = 7.5 \times 10^{-3}$$

$$H_2PO_4^- \rightleftharpoons H^+ + HPO_4^{--} \qquad K_2 = 6.2 \times 10^{-8}$$

$$HPO_4^{--} \rightleftharpoons H^+ + PO_4^{---} \qquad K_3 = 1.0 \times 10^{-12}$$

Thus H_3PO_4 is a stronger acid than $H_2PO_4^-$, which is stronger than HPO_4^{--}.

The ionization of bases and their equilibrium constants would follow a similar pattern: the larger the value for the ionization constant, the stronger the base.

Common ion effect. The ionic and molecular concentrations of 0.1 M $HC_2H_3O_2$ were evaluated in the last section.

$$K_a = \frac{[H^+] \times [C_2H_3O_2^-]}{[HC_2H_3O_2]} = \frac{0.00134 \times 0.00134}{0.09866} = 1.82 \times 10^{-5}$$

The ionization constant K_a is invariable at a given temperature and expresses a ratio of the product of the concentrations of the ions to the concentration of acetic acid molecules. If H^+ is added to the equilibrium mixture of acetic acid and its ions, the concentration of $C_2H_3O_2^-$ and concentration of $HC_2H_3O_2$ will have to change to maintain the constancy of K_a. Similarly, if more $C_2H_3O_2^-$ is added to the acetic acid and its ions in equilibrium, a change in concentration of H^+ ions and $HC_2H_3O_2$ molecules is necessary to keep K_a constant.

Problem 1. Assume that 8.2 g of solid $NaC_2H_3O_2$ is dissolved in 1 liter of 0.1 M $HC_2H_3O_2$. This is a 0.1 mole quantity of $NaC_2H_3O_2$, and the salt may be considered as ionized completely.

$$HC_2H_3O_2 \rightleftharpoons C_2H_3O_2^- + H^+$$

$$NaC_2H_3O_2 \longrightarrow C_2H_3O_2^- + Na^+$$

To evaluate the concentration of H^+ in such a solution, let x be the H^+ ion concentration. Note that now the acetate ion concentration will be increased above the H^+ ion concentration by 0.1 mole/liter. The molecular acetic acid concentration will be the initial value (0.1 M) decreased by the H^+ ion concentration; that is, $[H^+] = x$, $[C_2H_3O_2^-] = 0.1 + x$, and $[HC_2H_3O_2] = 0.1 - x$.

Then $$K_a = 1.82 \times 10^{-5} = \frac{x(0.1 + x)}{0.1 - x}$$

Since the value of x is small compared to 0.1, it may be neglected in the terms $0.1 + x$ and $0.1 - x$, and the equation reduces simply to

$$\frac{0.1x}{0.1} = 1.82 \times 10^{-5}$$

$$x = 1.82 \times 10^{-5} \text{ g ion/liter} = [H^+]$$

The concentration of H^+ ion, which was 0.00134 g ion/liter in 0.1 *M* acetic acid, was markedly decreased by the addition of the common acetate ion, thus rendering the $HC_2H_3O_2$ a still weaker acid. It would behave as if only 0.0182% ionized. By regulating the amount of $NaC_2H_3O_2$ added from none to enough to make the solution 0.1 *M* $NaC_2H_3O_2$, the acidity of the solution can be made to vary from the 1.34% to 0.0182% ionization; that is, the acidity is lessened.

In a similar manner the addition of a common ion to a weak base will repress the ionization of the base. For example,

$$NH_4OH \rightleftharpoons NH_4^+ + OH^-$$

$$NH_4I \longrightarrow NH_4^+ + I^-$$

Problem 2. (a) Calculate the OH^- ion concentration in 1.0 *M* NH_4OH; (b) calculate the OH^- ion concentration in a solution which is made by adding 14.5 g of NH_4I to a liter of 1.0 *M* NH_4OH. The ionization constant for NH_4OH is 1.8×10^{-5}.

$$NH_4OH \rightleftharpoons NH_4^+ + OH^-$$

(a) In pure 1.0 *M* NH_4OH, $[OH^-]$ would be equal to $[NH_4^+]$. Then, if $y = [OH^-]$,

$$K_b \text{ for } NH_4OH = 1.8 \times 10^{-5} = \frac{[NH_4^+] \times [OH^-]}{[NH_4OH]} = \frac{y^2}{1 - y}$$

Assume $1 - y = 1$, then $y = \sqrt{1.8 \times 10^{-5}} = 0.00424$

or 4.24×10^{-3} g ion OH^-/liter for pure 1 *M* NH_4OH.

(b) Molecular weight of $NH_4I = 145$, so that 14.5 g of NH_4I will furnish 0.1 mole or g ion of NH_4^+.

NH_4OH is ionized to a small extent, as indicated by the above equation and low K_b. As the additional NH_4^+ from NH_4I is added, K_b remains the same but the concentration of OH^- will lessen, and the concentration of NH_4OH molecules somewhat increases. Let x be the concentration of OH^- in a solution of 0.1 *M* NH_4I in 1.0 *M* NH_4OH.

Then $[NH_4OH] = 1 - x$ and $[NH_4^+] = (0.1 + x)$

$$\begin{array}{ccc} NH_4OH & \rightleftharpoons & NH_4^+ + OH^- \\ (1 - x) & & (0.1 + x)\; x \end{array}$$

$$K_b = 1.8 \times 10^{-5} = \frac{(0.1 + x)\, x}{1 - x}$$

Since x is small as compared with 0.1 and 1.0, it may be disregarded from the $[NH_4^+]$ and $[NH_4OH]$ terms with a small percentage of error, and the equation simplifies:

$$K_b = 1.8 \times 10^{-5} = \frac{0.1x}{1}$$

$x = 1.8 \times 10^{-4}$ g ion OH^-/liter for 1 *M* NH_4OH containing 0.1 *M* NH_4I

It is evident that addition of an ammonium salt to a solution of NH_4OH will markedly lower the OH^- concentration; in this problem it decreased from 4.24×10^{-3} g ion OH^-/liter to 1.8×10^{-4} g ion OH^-/liter.

Buffer solutions. The addition of a common ion (anion) to a weak acid not only lowers the acidity but also yields a solution which resists marked change of acidity. Assume that you have 1 liter of a solution containing 0.1 *M* $HC_2H_3O_2$ and 0.1 *M* $NaC_2H_3O_2$. If 0.1 mole of HCl gas is bubbled into this solution, will the full effect of the ionization of HCl to H^+ be exhibited? No, since the large excess of $C_2H_3O_2^-$ ion will use up much of the H^+ from the HCl to form slightly ionized $HC_2H_3O_2$.

$$H^+ + C_2H_3O_2^- \rightleftharpoons HC_2H_3O_2$$

It is also apparent that this solution would undergo little change in acidity on the addition of a base such as NaOH. The OH^- ions from the added NaOH will use up H^+, but more $HC_2H_3O_2$ will ionize. Similarly, a solution of NH_4Cl in NH_4OH resists marked increase in OH^- ion concentration. It may be stated that, generally, a solution of a weak acid and a salt of said acid, or a solution of a weak base and a salt of said base, will resist changes in acidity or basicity. Such solutions are termed *buffer solutions*.

Applications of buffer solutions are common in qualitative analysis to control acidity or basicity. They also find use in industrial processes. The *p*H of blood is kept within a small range by buffer action.

Hydrogen ion concentration, *p*H. Hydrogen ion concentration is a measure of the acidity of a solution and is usually expressed in moles per liter of solution.

In the application of science to the study of biological fluids, soil waters, and industrial solutions which are slightly acidic or basic, the *p*H* method for expressing hydrogen ion concentration is employed. We may best illustrate this method with examples: Let us consider the hydrogen ion concentration in solutions of hydrochloric acid. Hydrochloric acid is a strong acid and is therefore almost completely ionized. Assuming complete ionization, a solution of 0.01 *M* HCl will produce a 0.01 *M* concentration of hydrogen ions; a solution of 0.001 *M* HCl gives a hydrogen ion concentration of 0.001 *M*, and so on. These concentrations may be expressed as 10^{-2} and 10^{-3} ($0.01 = 10^{-2}$ and $0.001 = 10^{-3}$), respectively. The *p*H of the solution is taken as the exponent with its sign reversed.† Thus a solution of hydrogen ion concentration 10^{-2} would have a *p*H

* *p*H was originally proposed as a measure of hydrogen ion concentration. Although the term "hydrogen ion concentration" is still in common use, the ion which is actually responsible for acidity is the hydronium ion—that is, the hydrated H^+, H_3O^+.

† *p*H is defined as the negative logarithm of the hydrogen ion concentration, that is, $pH = -\log [H^+]$. Thus if $[H^+] = 10^{-2}$, $pH = -\log 10^{-2} = -(-2) = 2$.

of 2, a solution of hydrogen ion concentration 10^{-3} a pH of 3, and so on (see page 28). Note that a change of one unit in pH corresponds to 10 times the change in hydrogen ion concentration. The solution of pH 2 contains 10 times the concentration of hydrogen ions as the solution of pH 3. A study of Table 1 will further reveal the relationship between hydrogen ion concentration and pH. It may be noted from the table that the pH of the solution increases as the concentration of hydrogen ion decreases.

TABLE 1. *Hydrogen Ion Concentrations of Solutions of* HCl

Concentration of HCl *in moles or equivalents per liter*	*Concentration of hydrogen ion*	pH
0.1	0.1 $= 10^{-1}$	1
0.01	0.01 $= 10^{-2}$	2
0.001	0.001 $= 10^{-3}$	3
0.0001	0.0001 $= 10^{-4}$	4
0.00001	0.00001 $= 10^{-5}$	5
0.000001	0.000001 $= 10^{-6}$	6

The ionization of water. Pure water is a very poor conductor of electricity. This is emphasized by the fact that liquid mercury is about 25,000,000 times as good a conductor as water. The low conductivity of water may be attributed to its very slight ionization in accordance with the reaction

$$H_2O + H_2O \rightleftharpoons H_3O^+ + OH^-$$

or simply

$$H_2O \rightleftharpoons H^+ + OH^- \qquad (1)$$

Evidence indicates that only 1 molecule of water in 555,000,000 is ionized. This corresponds to 1 mole (1 g ion) of H^+ and also 1 mole (1 g ion) of OH^- per 10,000,000 liters of water. Consequently, in pure water,

$$[H^+] = [OH^-] = \frac{1 \text{ mole}}{10{,}000{,}000 \text{ liters}} = \frac{1}{10^7} = 1 \times 10^{-7} \text{ mole/liter}$$

The pH of pure water is 7, and, since equal numbers of H^+ and OH^- ions are present, this pH represents a neutral solution.

Applying the law of mass action to the above equilibrium,

$$K_i = \frac{[H^+] \times [OH^-]}{[H_2O]}$$

In pure water the molarity of H_2O is $1000 \div 18$ (mol. wt. H_2O) $= 55.5$, assuming that 1000 ml of water weighs 1000 g. In dilute aqueous solutions, the molarity of the H_2O is still about 55.5, and we may make the assumption that the concentration of H_2O in dilute solutions is a constant. Hence, $K_i \times [H_2O]$ is a constant and may be represented by another constant K_w. The value of K_w at room temperature is 10^{-14}. This value is the product of the concentrations of hydrogen ion and hydroxyl ion in any water solution.

$$K_w = [H^+] \times [OH^-]$$

$$K_w = (1 \times 10^{-7})(1 \times 10^{-7}) = 1 \times 10^{-14}$$

Let us suppose that an acid is added to water until the hydrogen ion concentration is 10^{-3} (pH = 3). The increase in hydrogen ion concentration displaces the equilibrium (1) above to the left and results in a decrease in the concentration of OH^-. The value of the concentration of OH^- in the solution must be 10^{-11}, since $10^{-3} \times 10^{-11} = 10^{-14}$. Students frequently think of an acid solution as containing hydrogen ions but no hydroxyl ions. It is evident from the above that any aqueous solution of an acid also contains hydroxyl ions. Likewise, any basic or alkaline solution contains not only hydroxyl ions but hydrogen ions as well. A solution in which the concentration of OH^- is 10^{-4} must have a hydrogen ion concentration of 10^{-10}. The pH of this latter solution would be 10, and since $[OH^-]$ is greater than $[H^+]$, the solution is basic. Summarizing, a solution of pH 7 is neutral, a solution of pH less than 7 is acid, and one of pH greater than 7 is alkaline.

$$0 \xleftarrow{\text{Increasing acidity}} \underset{\text{Neutral}}{7} \xrightarrow{\text{Increasing basicity}} 14$$

Since concentrations of hydrogen ion in solutions are seldom exact integral powers of 10, pH values may be numbers such as 3.42, 8.74, and so on (see page 28). A careful control of pH is often important in qualitative analysis.

7

Use of logarithms and calculations of pH and pOH

The use of logarithms. The logarithm of any number is the power to which 10 must be raised to equal the number.

Examples.

10,000	$= 10^4$	logarithm is	4.0
100	$= 10^2$		2.0
1	$= 10^0$		0.0
0.1	$= 10^{-1}$		-1.0 or $(9.0000 - 10)$
0.001	$= 10^{-3}$		-3.0 or $(7.0000 - 10)$

For a number such as 63, which can not be expressed as an integral power of 10, one understands that it is a number between 10 (10^1) and 100 (10^2), and thus its logarithm must be between 1 and 2 (actually 1.7993, as a logarithm table indicates). Every logarithm has two parts: an integer called the *characteristic*, and a decimal (usually endless) called the *mantissa*. In the example above, the logarithm of 63 is 1.7993. The integer 1 to the left of the decimal point is the characteristic, and .7993 is the mantissa.

The characteristic of the logarithm is determined by the following rules: (1) For numbers greater than 1, the characteristic is one less than the number of digits before the decimal point; for example, for the number 456. with three digits, the characteristic of the logarithm will be 2; for 35. the characteristic will be 1. (2) For numbers less than 1, the characteristic will be a negative integer one more than the number of zeros to the right of the decimal point in the number; for example, for the number 0.735 the characteristic is -1; for the number 0.00064 the characteristic is -4. Since the mantissa is always a positive value, characteristics with a negative value are usually expressed as a number minus 10; thus a characteristic of -1 will be expressed as $9.0000 - 10$; -3 will be expressed as $7.0000 - 10$. The mantissa of a number is obtained from a logarithm table which may be found in all chemistry, physics, or mathematics handbooks. (A four-place table will be found in the Appendix of this manual.)

Examples. [Consult the logarithm table and verify the following logarithms (logs). One mentally supplies the characteristic and looks up the mantissa in the table.]

$$\log 627.2 = 2.7974$$
$$\log 0.00423 = 7.6263 - 10$$
$$\log 87210.0 = 4.9406$$
$$\log 0.0000297 = 5.4728 - 10$$

If the logarithm of a number is known, the number (called the *antilog*) may be obtained by consulting a logarithm table. For example, find the number corresponding to the logarithm 2.8156. Looking up the mantissa .8156 in the four-place log table gives the digits 654. Since the characteristic is 2, the actual number must be between 100 (10^2) and 1000 (10^3). Therefore, the number is 654.

The number of significant figures is limited by the accuracy of the log tables—four place, five place, seven place, and so on. The mantissa gives only a set of digits, and the characteristic is used to locate the decimal point. In the above, if the characteristic had been 5 rather than 2, we would arrive again at the digits 654 from the four-place log table, but the decimal would have to be placed so that the number would be 654000. The last three digits are not known with accuracy, and so the number is better expressed as 6.54×10^5.

Operations with logarithms.

MULTIPLICATION. In multiplying numbers, the logarithms of the numbers are *added*, and their sum is the logarithm of the product.

Example 1. $627.2 \times 87210 = ?$

$$\begin{aligned} \log 627.2 &= 2.7974 \\ \log 87210 &= \underline{4.9406} \\ \text{log of product} &= 7.7380 \\ \text{antilog} &= 54700000.\ \text{or}\ 5.47 \times 10^7\ \text{(answer)} \end{aligned}$$

Example 2. $627.2 \times 0.00423 = ?$

$$\begin{aligned} \log 627.2 &= 2.7974 \\ \log 0.00423 &= \underline{7.6263 - 10} \\ \text{log product} &= 10.4237 - 10 \\ \text{antilog} &= 2.653\ \text{(answer)} \end{aligned}$$

DIVISION. In dividing numbers, the logarithms of the numbers are *subtracted* to obtain the log of the quotient.

Example 1. $627.2 \div 87210 = ?$

$$\begin{aligned} \log 627.2 &= 12.7974 - 10\ \text{(characteristic 2)} \\ \log 87210 &= \underline{4.9406} \\ \text{log quotient} &= 7.7568 - 10 \\ \text{antilog} &= 0.00571\ \text{or}\ 5.71 \times 10^{-3}\ \text{(answer)} \end{aligned}$$

Example 2. $0.00423 \times 0.0000297 = ?$

$$\begin{aligned} \log 0.00423 &= 7.6263 - 10 \\ \log 0.0000297 &= \underline{5.4728 - 10} \\ \text{log quotient} &= 2.1535 \\ \text{antilog} &= 1.42 \times 10^2 \end{aligned}$$

ROOTS. To take the root of a number, divide the logarithm of the number by the root and obtain the logarithm of the answer.

Example 1. $\sqrt[2]{627.2} = ?$

$$\log 627.2 = 2.7974$$
$$2.7974 \div 2 = 1.3987$$
$$\text{antilog of } 1.3987 = 25.05 \text{ (answer)}$$

Example 2. $\sqrt[3]{627.2} = ?$

$$\log 627.2 = 2.7974$$
$$2.7974 \div 3 = 0.9324$$
$$\text{antilog of } 0.9324 = 8.559 \text{ (answer)}$$

POWERS. To raise a number to a power, multiply the logarithm of the number by the power and obtain the logarithm of the answer.

Example 1. $(627.2)^2 = ?$

$$\log 627.2 = 2.7974$$
$$\log \text{ of } (627.2)^2 = 2 \times 2.7974 = 5.5948$$
$$\text{antilog} = 3.934 \times 10^5 \text{ (answer)}$$

Example 2. $(627.2)^4 = ?$

$$\log 627.2 = 2.7974$$
$$\log \text{ of } (627.2)^4 = 4 \times 2.7974 = 11.1896$$
$$\text{antilog} = 1.548 \times 10^{11} \text{ (answer)}$$

Logarithmic calculations of *p*H and *p*OH.

*p*H is defined as the negative logarithm of the hydrogen ion concentration or the logarithm of the reciprocal of the hydrogen ion concentration. These are equal; that is,

$$p\text{H} = -\log [\text{H}^+] = \log \frac{1}{[\text{H}^+]} = \log 1 - \log [\text{H}^+] = 0 - \log [\text{H}^+]$$

Similarly,

$$p\text{OH} = -\log [\text{OH}^-] = \log \frac{1}{[\text{OH}^-]} = \log 1 - \log [\text{OH}^-] = 0 - \log [\text{OH}^-]$$

Study these calculations:

Problem 1. A 0.1 *M* acetic acid solution has a hydrogen ion concentration of 0.00134 moles/liter. What is its *p*H?

ALTERNATE SOLUTION

$$pH = \log \frac{1}{[H^+]} = \log 1 - \log 0.00134$$

$$\log 1 = 0 \text{ or } 10.0000 - 10$$
$$\log 0.00134 = 7.1271 - 10$$
$$pH = 2.8729$$

$$pH = -(\log 1.34 \times 10^{-3})$$

or

$$-pH = \log 1.34 + \log 10^{-3}$$
$$-pH = 0.1271 - 3$$
$$pH = 2.8729$$

Problem 2. A 1 *M* NH_4OH solution buffered with 0.1 *M* NH_4I has an OH^- concentration of 1.8×10^{-4} moles/liter. What is its *p*OH? Its *p*H?

$$pOH = \log \frac{1}{[OH^-]} = \log \frac{1}{1.8 \times 10^{-4}} = \log 1 - \log 1.8 \times 10^{-4}$$

$$\log 1 = 10.0000 - 10$$
$$\log 1.8 \times 10^{-4} = 6.2553 - 10$$
$$pOH = 3.7447$$

Since $$K_w = [H^+] \times [OH^-] = 1 \times 10^{-14}$$

then $$pH + pOH = 14$$
$$pH = 14 - pOH$$
$$pH = 14 - 3.7447$$
$$pH = 10.2553$$

Problem 3. A 0.1 *M* solution of H_3PO_4 (0.3 *N*) is apparently 27% ionized. What is the appropriate *p*H of the solution?

$$H_3PO_4 \rightleftharpoons H^+ + H_2PO_4^-$$
$$[H^+] = 0.1 \times 0.27 = 0.027$$
$$pH = \log 1 - \log 0.027$$
$$\log 1 = 10.0000 - 10$$
$$\log 0.027 = 8.4314 - 10$$
$$pH = 1.5686$$

Subsequent ionization of $H_2PO_4^-$ adds a little more H^+ (slightly lower *p*H), but this will not be considered here.

8
Solubility product principle

Ions of a slightly soluble salt. The principles of ionic equilibria apply equally, as we have said, to the ionization of a weak acid or base and to the presence of ions in a solution of a slightly soluble salt. Our discussion will be limited to slightly soluble salts for two reasons: certain complicating factors render the law of mass action more or less invalid for concentrated solutions of ions such as exist in solutions of quite soluble salts; further, it is the dilute solution of ions in contact with a fairly insoluble precipitate which is most frequently important in qualitative analytical separations.

If solid AgCl is shaken with water, Ag^+ and Cl^- from the crystal lattice of the solid pass into solution until the solution becomes saturated and an equilibrium exists between the ions in solution and the solid:

$$\underset{\text{solid}}{AgCl} \rightleftharpoons Ag^+ + Cl^-$$

The amount of AgCl dissolved depends on the temperature; at a given temperature the solubility of a substance is a constant. Applying the law of mass action to the above equilibrium, we may write:

$$K_i = \frac{[Ag^+] \times [Cl^-]}{[AgCl]}$$

It will be observed that the concentration of solid AgCl, as with any solid, does not vary (the density of a given solid is constant), and we may assume that the term in the denominator is a constant. The equation then simplifies to

$$K_{sp} = [Ag^+] \times [Cl^-]$$

where K_{sp} is the equilibrium constant given the special name *solubility product*. The product of the concentrations of ions in a saturated solution of a relatively insoluble salt at a given temperature is a constant. This is known as the *solubility product principle.*

The numerical value for K_{sp} can readily be calculated if the concentrations of ions in the saturated solution are known. If a saturated solution of AgCl is prepared at 25° C, the solubility by analysis is found to be 1.1×10^{-5} mole/liter. AgCl is completely ionized, hence each formula weight of AgCl will give 1 g ion of Ag^+ and 1 g ion of Cl^- in solution; consequently, the concentration of Ag^+ will be 1.1×10^{-5} g ion/liter and the concentration of Cl^- will be 1.1×10^{-5} g ion/liter. Substituting these concentrations in the solubility product formulation,

$$K_{sp} = [Ag^+] \times [Cl^-]$$

$$K_{sp} = (1.1 \times 10^{-5})(1.1 \times 10^{-5}) = 1.2 \times 10^{-10}$$

The value for K_{sp} is a measure of the solubility of a substance; the larger the value of K_{sp}, the greater must be the concentrations of ions, and hence the greater the solubility of the substance. The smaller the value of K_{sp}, the more *insoluble* the substance.

In many cases more than two ions are produced in the ionization of a compound. Where more than two ions are involved on one side of the equation, the coefficients of the ions must be taken into account. The following problem illustrates the point. Calculate the solubility product for $Ca_3(PO_4)_2$, given the fact that 2.5×10^{-2} g of it dissolves in 1 liter water at 20° C.

$$\text{mol. wt.} = 310$$

$$\text{solubility} = 2.5 \times 10^{-2}/310 = 8.06 \times 10^{-5} \text{ mole/liter}$$

$$Ca_3(PO_4)_2 \rightleftharpoons 3\ Ca^{++} + 2\ PO_4^{3-}$$

$$K_{sp} = [Ca^{++}]^3 \times [PO_4^{3-}]^2$$

Each $Ca_3(PO_4)_2$ yields 3 Ca^{++} and 2 PO_4^{---}; hence, 8.06×10^{-5} mole of $Ca_3(PO_4)_2$ yields $3 \times 8.06 \times 10^{-5}$ g ion Ca^{++} and $2 \times 8.06 \times 10^{-5}$ g ion PO_4^{3-}.

$$\begin{aligned} K_{sp} &= (3 \times 8.06 \times 10^{-5})^3(2 \times 8.06 \times 10^{-5})^2 \\ &= (24.2 \times 10^{-5})^3(16.1 \times 10^{-5})^2 \\ &= (1.4 \times 10^{-11})(2.59 \times 10^{-8}) \\ &= 3.6 \times 10^{-19} \end{aligned}$$

The very low solubility product for calcium phosphate is an index of its low solubility. (A table of solubility products is in the Appendix.)

Application of solubility product principle to formation of precipitates. The formation of a precipitate depends first on the formation of a saturated solution; only when the solubility of the substance is exceeded can a precipitate form. K_{sp} is directly related to the solubility of a substance since it is equal to the product of ion concentrations in a saturated solution. Any solution in which the product of the ion concentrations exceeds K_{sp} for the substance is supersaturated, and precipitation will occur until the solution is saturated, in which case the product of ion concentrations attains the value of K_{sp}. Any solution in which the ion product is less than K_{sp} is unsaturated and is capable of dissolving more of the solute.

When ions are brought together in solution and the product of the concentrations of the ions involved is greater than the solubility product, precipitation takes place.

In the case of AgCl, when the product of $[Ag^+]$ and $[Cl^-]$ exceeds 1.2×10^{-10}, its K_{sp}, the salt will precipitate. In pure water solution of AgCl, the concentrations of Ag^+ and Cl^- are the same, that is, 1.1×10^{-5} mole/liter. It is not necessary that the concentrations of the ions have equivalent values to bring about precipitation, so long as the product of these ions concentrations exceeds K_{sp}.

Suppose we know the concentration of Ag^+ in a certain solution to be 1×10^{-7} g ion/liter. We can calculate the concentration of Cl^- necessary to bring about precipitation. Substituting in the equation

$$K_{sp} = [Ag^+] \times [Cl^-] = 1.2 \times 10^{-10}$$

and solving for $[Cl^-]$

$$[Cl^-] = \frac{1.2 \times 10^{-10}}{1 \times 10^{-7}} = 1.2 \times 10^{-3}$$

This concentration of Cl^- will just form a saturated solution, but any value of $[Cl^-]$ larger than 1.2×10^{-3} will start precipitation of AgCl.

It was shown above that the solubility of AgCl in pure water was 1.1×10^{-5} mole/liter, while the solubility in a solution in which $[Cl^-] = 1.2 \times 10^{-3}$ is only 1×10^{-7}; in other words, by using an excess of Cl^- the solubility of the salt has been decreased nearly a hundredfold. In the precipitation of ions from solution in qualitative and quantitative analysis, a large excess of the precipitating ion is used to reduce markedly the solubility of the substance and cause it to precipitate. Study the following illustrations:

Example 1. Assume a solution is made 0.001 *M* with respect to $AgNO_3$ and simultaneously made 0.001 *M* with respect to NaCl. Will precipitation occur? The product of the concentration of Ag^+ and that of Cl^- is $(0.001)^2$ or 1×10^{-6}. This is greater than K_{sp} for AgCl (1.2×10^{-10}), and a precipitate will be formed.

Example 2. Positive and negative ions need not be brought together in equal amounts, and so let us analyze the case of having a 0.1 g ion Ag^+ solution and adding a chloride to it to make the Cl^- concentration 1×10^{-11}. Will precipitation take place? The ion product $0.1 \times 1 \times 10^{-11} = 1 \times 10^{-12}$ is less than K_{sp} for AgCl of 1.2×10^{-10}, and no precipitation results.

Example 3. K_{sp} for $PbSO_4$ is 2.3×10^{-8}. Given a solution containing 1×10^{-5} g ion Pb^{++}/liter, what is the maximum concentration of SO_4^{--} that may be present without the formation of a precipitate of $PbSO_4$?

Let

$$x = [SO_4^{--}]$$

$$K_{sp} = 2.3 \times 10^{-8} = [Pb^{++}] \times [SO_4^{--}]$$

$$2.3 \times 10^{-8} = 1 \times 10^{-5} \times x$$

$$x = 2.3 \times 10^{-3} = [SO_4^{--}]$$

to yield a saturated solution.

Example 4. A solution is 0.1 *M* with respect to $MgCl_2$ and NH_4OH. Will precipitation take place and, if so, what concentration of Mg^{++} remains in solution?

$$K_{sp} \text{ of } Mg(OH)_2 = [Mg^{++}] \times [OH^-]^2 = 3.2 \times 10^{-11}$$

The concentration of Mg^{++} is 0.1 g ion/liter, since the $MgCl_2$ may be assumed to be 100% as ions. The concentration of OH^- from the 0.1 *M* NH_4OH must be calculated, and this may be done from the ionization constant for NH_4OH, thus:

$$K_b \text{ for } NH_4OH = 1.8 \times 10^{-5} = \frac{[NH_4^+][OH^-]}{[NH_4OH]}$$

Let $x = [OH^-]$, then $[NH_4^+] = x$, and $[NH_4OH] = 0.1 - x$. Substituting,

$$\frac{(x)(x)}{0.1 - x} = 1.8 \times 10^{-5}$$

In the term $(0.1 - x)$, x is small compared with 0.1, and so assume that $0.1 - x = 0.1$. Then

$$\frac{x^2}{0.1} = 1.8 \times 10^{-5}$$

$$x^2 = 1.8 \times 10^{-6}$$

$$x = 1.34 \times 10^{-3} = [OH^-]$$

Instantaneous concentrations on mixing are 0.1 g ion/liter Mg^{++} and 1.34×10^{-3} g ion/liter OH^-. Ionic product of same:

$$0.1 \times (1.34 \times 10^{-3})^2 = 1.8 \times 10^{-7}$$

This is greater than K_{sp}, so precipitation will take place.

$$K_{sp} \text{ of } Mg(OH)_2 = 3.2 \times 10^{-11} = [Mg^{++}] \times (1.34 \times 10^{-3})^2$$

$$[Mg^{++}] \text{ left in solution} = \frac{3.2 \times 10^{-11}}{(1.34 \times 10^{-3})^2} = 1.78 \times 10^{-5} \text{ g ion/liter}$$

Example 5. Calculate the concentration of lead ions and of chloride ions in a saturated solution of $PbCl_2$. Express the answer in both gram ions per liter and grams per liter. The solubility product of $PbCl_2$ is 2.4×10^{-4} at 20° C.

$$PbCl_2 \rightleftharpoons Pb^{++} + 2\ Cl^-$$

$$K_{sp} = [Pb^{++}] \times [Cl^-]^2 = 2.4 \times 10^{-4}$$

Let $x = [Pb^{++}]$ and $y = [Cl^-]$.

$$x \times y^2 = 2.4 \times 10^{-4}$$

but $y = 2x$, since, from $PbCl_2$, $[Cl^-]$ is twice $[Pb^{++}]$.

Therefore $$x \times (2x)^2 = 4x^3 = 2.4 \times 10^{-4}$$

$$x^3 = 0.6 \times 10^{-4}$$

$$x = 0.0391 = [Pb^{++}]$$

$$[Pb^{++}] = 0.0391 \text{ g ion/liter} \times 207 = 8.09 \text{ g/liter}$$

$$[Cl^-] = 0.0782 \text{ g ion/liter} \times 35.5 = 2.78 \text{ g/liter}$$

The dissolving of precipitates. Just as a precipitate is formed when the product of ion concentrations exceeds K_{sp}, so a precipitate must dissolve when the ion concentrations are decreased sufficiently to give an ion product less than K_{sp}. Anything which will lower the concentrations of ions in the solution will disturb the equilibrium with the undissolved solid, and more solid will pass into solution. This process will continue until all the solid has passed into solution if the concentrations of ions are continuously lowered.

> **When the product of the concentrations of the ions in solution is less than the solubility product, undissolved solid composed of those ions will dissolve.**

Accordingly, as long as the product of ionic concentrations above an undissolved salt is less than the solubility product, salt will continue to dissolve.

The ionic concentration product may be kept low in the following ways:

1. DILUTION. Dilution lessens ionic concentrations. It stands to reason that more salt dissolves as more solvent is added.

2. BY DESTROYING THE ANION OF THE SALT. $CaCO_3$ dissolves in dilute acid:

$$\begin{array}{l} CaCO_3 \rightleftharpoons Ca^{++} + \boxed{CO_3^{--}} \\ 2\,HCl \rightarrow 2\,Cl^- + \boxed{2\,H^+} \end{array}$$

$$\downarrow$$

$$H_2CO_3 \searrow CO_2\uparrow + H_2O$$

3. REMOVING ONE OF THE IONS OF THE SALT FROM SOLUTION BY FORMATION OF A SLIGHTLY IONIZED OR INSOLUBLE SUBSTANCE. (a) Lead sulfate will dissolve in ammonium acetate, as indicated by the equations

$$\begin{array}{l} PbSO_4 \rightleftharpoons \boxed{Pb^{++}} + SO_4^{--} \\ 2\,NH_4C_2H_3O_2 \rightarrow \boxed{2\,C_2H_3O_2^-} + 2\,NH_4^+ \end{array}$$

$$\downarrow$$

$Pb(C_2H_3O_2)_2$ (an example of one of the few slightly ionized salts)

(b) Zinc ion will enter solution if ZnS powder is agitated in $HgCl_2$ solution:

$$\begin{array}{l} ZnS \rightleftharpoons Zn^{++} + \boxed{S^{--}} \\ HgCl_2 \rightarrow 2\,Cl^- + \boxed{Hg^{++}} \end{array}$$

$$\downarrow$$

$$HgS\downarrow$$

Application of the solubility product principle to the precipitation of cations of Group II. The precipitation of the sulfides of Group II is brought about by means of H_2S (from thioacetamide), a weak acid which ionizes in two stages:

$$H_2S \rightleftharpoons H^+ + HS^- \qquad K_1 = \frac{[H^+] \times [HS^-]}{[H_2S]} = 1.1 \times 10^{-7}$$

$$HS^- \rightleftharpoons H^+ + S^{--} \qquad K_2 = \frac{[H^+] \times [S^{--}]}{[HS^-]} = 1 \times 10^{-15}$$

Combining the primary and secondary ionization constants,

$$K_1K_2 = K_a = \frac{[H^+]^2 \times [S^{--}]}{[H_2S]} = 1.1 \times 10^{-22}$$

In a saturated solution of H_2S at 25° C the solubility of H_2S is 0.1 mole/liter, and the concentration of sulfide ion by experiment has been shown to be 1.2×10^{-15}. Since the concentration of H_2S remains essentially a constant (0.1), we may write

$$K_a \times [H_2S] = (1.1 \times 10^{-22})(0.1) = [H^+]^2 \times [S^{--}] = 1.1 \times 10^{-23} \qquad (1)$$

The concentration of H^+ in the water solution of H_2S may be calculated thus:

Substituting 1.2×10^{-15} for $[S^{--}]$:

$$[H^+]^2 \times 1.2 \times 10^{-15} = 1.1 \times 10^{-23}$$

$$[H^+]^2 = \frac{1.1 \times 10^{-23}}{1.2 \times 10^{-15}} = 91.7 \times 10^{-10}$$

$$[H^+] = \sqrt{91.7 \times 10^{-10}} = 9.5 \times 10^{-5} \text{ mole/liter} \quad (p\text{H near } 4)$$

Since the product of the ion concentrations of H_2S is always 1.1×10^{-23}, it is evident that the sulfide ion concentration may be made greater than 1.2×10^{-15} or less than 1.2×10^{-15} by simply decreasing or increasing the H^+ concentration. If an acid is added to the saturated H_2S solution, the H^+ ion concentration is increased, and the concentration of S^{--} must be decreased in order that the product of the ion concentrations give 1.1×10^{-23}. On the other hand, the addition of a base (which takes up H^+) would decrease the concentration of H^+, and the S^{--} concentration would increase in order to satisfy the equilibrium value, namely 1.1×10^{-23}. Consequently, by regulation of the acidity or basicity of the solution, the sulfide ion concentration may be carefully controlled.

The regulation of $[S^{--}]$ is the basis for the separation of Group II metals. The solubility product constants of common metals sulfides vary from 4.0×10^{-53} for HgS to 1.4×10^{-15} for MnS. It should be evident, then, that a much larger S^{--} concentration is necessary to precipitate the latter than the former. When H_2S is added to solutions of Hg^{++}, Cu^{++}, Cd^{++}, Pb^{++}, Zn^{++}, Fe^{++}, or Mn^{++}, the sulfides will precipitate in the order of increasing value of K_{sp}; that is, HgS will precipitate first, CuS next, CdS next, and so forth, and MnS last. If the sulfide ion concentration is properly controlled, only the sulfides in Group II—that is, HgS, CuS, CdS, and PbS—will precipitate. It has been determined experimentally that *the concentration of* H^+ *which will effect this separa-*

tion is about 0.3 *M*. If we substitute this value for the H^+ concentration in (1) above:

$$(0.3)^2 \times [S^{--}] = 1.1 \times 10^{-23}$$

And, solving for $[S^{--}]$,

$$[S^{--}] = 1.2 \times 10^{-22}$$

Now let us check the precipitation of several cations as sulfides in this concentration of sulfide ion. We shall determine the concentrations of Hg^{++}, Pb^{++}, and Fe^{++} which would have to be present to give the corresponding sulfide precipitates in a solution containing sulfide ion at a concentration of 1.2×10^{-22} g ion/liter.

Problem 1. What concentration of Hg^{++} must be exceeded to give a precipitate of HgS?

Since K_{sp} for HgS $= 4.0 \times 10^{-53}$,

$$[Hg^{++}] \times [S^{--}] = 4.0 \times 10^{-53}$$

Substituting the above value for $[S^{--}]$, and solving for $[Hg^{++}]$:

$$[Hg^{++}] = \frac{4.0 \times 10^{-53}}{1.2 \times 10^{-22}} = 3.3 \times 10^{-31}$$

which is the $[Hg^{++}]$ necessary to just form a saturated solution of HgS. Any value above this would cause the precipitation of HgS. Since this is an extremely small concentration of Hg^{++} (it is normally much larger than this in any sample for analysis containing mercury), HgS will precipitate when H_2S is added to the solution.

Problem 2. What concentration of Pb^{++} will form a saturated solution with H_2S in the presence of 0.3 g ion concentration of H^+?

PbS has the largest K_{sp} of the sulfides which precipitate in Group II.

$$K_{sp} \text{ for PbS} = 4.2 \times 10^{-28}$$

Then

$$[Pb^{++}] \times [S^{--}] = 4.2 \times 10^{-28}$$

If $[S^{--}] = 1.2 \times 10^{-22}$, we may substitute this value in the above and solve for $[Pb^{++}]$.

$$[Pb^{++}] = \frac{4.2 \times 10^{-28}}{1.2 \times 10^{-22}} = 3.5 \times 10^{-6}$$

Any concentration of Pb^{++} above 3.5×10^{-6} will cause precipitation of PbS. Since this is a smaller concentration of Pb^{++} than is ordinarily met with in analysis, PbS will precipitate in Group II when H_2S is added.

Problem 3. FeS is typical of the sulfides not precipitating in Group II but precipitating in Group III. Calculate the concentration of Fe^{++} which would have to be present to precipitate FeS in 0.3 g ion concentration of H^+.

$$K_{sp} \text{ for FeS} = 1.5 \times 10^{-19}$$

$$[Fe^{++}][S^{--}] = 1.5 \times 10^{-19}$$

$$[Fe^{++}] = \frac{1.5 \times 10^{-19}}{1.2 \times 10^{-22}} = 1.25 \times 10^{3} = 1250 \text{ g ion/liter}$$

It is not possible to obtain a 1250 M solution of Fe^{++}; hence FeS cannot precipitate in 0.3 M acid solutions.

Thus we see that a H^+ concentration of 0.3 M will effect a separation of the members of Group II from Group III because in this solution the S^{--} concentration is great enough to exceed the solubility product constants for the sulfides of Group II but not of Group III. Since $[S^{--}]$ is automatically controlled by $[H^+]$, it is necessary that the latter be carefully adjusted to the proper value – 0.3 g ion – in order to effect a clean-cut separation. The division of the sulfides between PbS and FeS has been made only because it is convenient to test for Pb^{++} in Group II and to test for Fe^{++} in a later group. The point of division could be shifted, of course, by changing the H^+ concentration of the solution.

The marked differences in K_{sp} for metal chlorides, sulfates, carbonates, chromates, and so forth is the basis for various metal ion separations in the procedural part of this manual.

9
Valence and oxidation number

Since a knowledge of valence and a thorough understanding of the concept of oxidation number are fundamental to the writing and balancing of equations, it is recommended that these items be reviewed. A table of standard electrovalences and a brief review of oxidation number is given below.

TABLE 2. *Electrovalences of Elements and Radicals*

+I	+II	+III	+IV
H^+ or H_3O^+ (hydronium) Na^+ K^+ Ag^+ NH_4^+ (ammonium) Cu^+ (cuprous)† Hg_2^{++} (mercurous)	Ca^{++} Mg^{++} Ba^{++} Cd^{++} Zn^{++} Pb^{++} Ni^{++} Cr^{++} (chromous) Fe^{++} (ferrous) Cu^{++} (cupric) Sn^{++} (stannous) Mn^{++} (manganous) Sr^{++} Co^{++} Hg^{++} (mercuric)	Al^{3+} Fe^{3+} (ferric) Cr^{3+} (chromic) As^{3+} Sb^{3+} Bi^{3+} La^{3+}	Sn^{4+} (stannic)
−I	**−II**	**−III**	**−IV**
Cl Cl^- (chloride) Br^- (bromide) I^- (iodide) F^- (fluoride) OH^- (hydroxide) $C_2H_3O_2^-$ (acetate) ClO_3^- (chlorate) NO_3^- (nitrate) CN^- (cyanide) NO_2^- (nitrite)	O^{--} (oxide) S^{--} (sulfide) SO_3^{--} (sulfite) SO_4^{--} (sulfate) CO_3^{--} (carbonate) CrO_4^{--} (chromate) $C_2O_4^{--}$ (oxalate) SiO_3^{--} (metasilicate)	PO_4^{3-} (phosphate) AsO_4^{3-} (arsenate) BO_3^{3-} (borate)	SiO_4^{4-} (silicate)

† The lower valence of a metal in a compound has long been designated with the suffix *-ous*; the higher valence with the suffix *-ic*, for example, Cu^+ (cuprous) and Cu^{++} (cupric) ions. Salts containing these ions—for example, Cu_2SO_4 and $CuSO_4$—would be cuprous sulfate and cupric sulfate, respectively. A more recent method is to show the actual valence or oxidation number of the metal with Roman numerals; thus, Cu_2SO_4 is copper (I) sulfate; $CuSO_4$ is copper (II) sulfate. Similarly, $CrCl_2$ is chromium (II) chloride and $CrCl_3$ is chromium (III) chloride.

Oxidation number. Since all atomic linkages are believed to involve electrical attraction, although the electron sharing bonds cloud the simple picture, an *apparent charge* to each atom in combination may be assigned. It may be thought of as what the charge on each atom of a compound would be if all the elements were electrovalent. Accordingly, chemists have agreed to use a term "oxidation number," which is determined from the set of rather arbitrary rules listed in Table 3.

TABLE 3. *Rules for Determining Oxidation Numbers*

Oxidation number of	*Magnitude of, or manner of determining*
Uncombined elements	0
Hydrogen	+1
Oxygen	−2
Ions alone or in electrovalent compounds	+ or − according to their electrovalent charge.
Elements in a binary covalent compound	The charge on each atom if the shared electrons are all assigned to the more electronegative of the atoms
The central element in a ternary compound	The apparent charge on the atom which will give the compound a neutral character when positive and negative oxidation numbers have been assigned to all other elements in the compound; the algebraic sum of the apparent charges in a compound is zero

Let us apply these rules to the determination of some oxidation numbers.

Carbon in the free state has an oxidation number of 0. In CO its oxidation state will be represented by the number +2, since oxygen is assigned a number −2. In CO_2 the oxidation number of carbon is +4. In CCl_4 the oxidation number of carbon is +4; in CH_4 each hydrogen is +1 in oxidation number, and carbon is −4.

In NaCl, an ionic compound, the oxidation number of sodium is +1 and of chlorine −1. In $AlBr_3$ the oxidation number of aluminum is +3, of each bromine −1. In the ionic compound $(NH_4)_2SO_4$ the oxidation number of each NH_4^+ is +1; of SO_4^{--}, −2. For an ion such as PO_4^{3-} the oxidation number of P must be +5, since the latter added to a total of −8 from the 4 oxygen atoms would leave a net charge of −3, which is the charge of the PO_4^{3-} ion. Study these examples:

COMPOUND	OXIDATION NUMBER
H_2SO_4	sulfur, +6
$H_4P_2O_7$	phosphorus, +5
Al_2S_3	aluminum, +3; sulfur, −2
MgO	magnesium, +2
$NaMnO_4$	sodium, +1; manganese, +7
$Na_2Cr_2O_7$	sodium, +1; chromium, +6
$HBrO_3$	bromine, +5
SiO_4^{4-}	silicon, +4
AsO_4^{3-}	arsenic, +5

10
Chemical changes in qualitative analysis

Ionic union. This type of chemical change (sometimes called *double decomposition* reactions) is by far the most common in qualitative analysis. The reactions result from a combination of the ions in aqueous solutions of acids, bases, and salts. Such reactions may be represented with molecular or ionic equations. In developing an ionic equation, students may find it easier in many cases to write a molecular equation first. Examples from qualitative analysis of this type of equation are:

$$PbCl_2 + H_2SO_4 \rightarrow PbSO_4\downarrow + 2\ HCl$$

$$Pb^{++} + 2\ Cl^- + 2\ H^+ + SO_4^{--} \rightarrow PbSO_4\downarrow + 2\ H^+ + 2\ Cl^-$$

or simply

$$Pb^{++} + SO_4^{--} \rightarrow PbSO_4\downarrow$$

$$CuCl_2 + H_2S \rightarrow CuS\downarrow + 2\ HCl$$

or simply

$$Cu^{++} + S^{--} \rightarrow CuS\downarrow$$

Oxidation-reduction. Any reaction in which a change of oxidation number occurs is classified as oxidation-reduction. In writing and balancing equations for these reactions, three different representation methods are used: (1) molecular equation, (2) ionic equation, (3) ion-electron method.

It is assumed that the student of qualitative analysis has studied oxidation-reduction, but we will summarize the concept here as a review. Consider the reaction

$$Fe^0 + S^0 \rightarrow Fe^{++}S^{--}$$

Fundamentally, 1 iron atom gives 2 electrons to 1 sulfur atom to form ferrous and sulfide ions, which place themselves in a solid crystal lattice. Iron is said to be oxidized; sulfur is the oxidizing agent. Sulfur may be said to be reduced; iron is the reducing agent. Iron gains 2 in oxidation number, and sulfur loses 2 in oxidation number. Thus:

Fe loses electrons, gains in oxidation number, is oxidized, and is the reducing agent.
S gains electrons, loses in oxidation number, is reduced, and is the oxidizing agent.

In oxidation-reduction processes involving ions, electrons are very definitely transferred from atom to atom. With covalent substances such a transfer does not occur. It is likely, however, that there is at least a displacement of electrons (always toward the more electronegative element). Therefore, in a sense, electrons are lost from one element and gained by another, and we may regard all oxidation-reduction processes as involving a gain and loss of electrons.

By balancing the gain and loss of electrons, we can obtain the ratio of the oxidizing agent to the reducing agent. From this information, the equation is easily balanced. Of course, the same information relative to the ratio may be obtained by considering a change in oxidation number. The following examples will make this clear.

MOLECULAR EQUATIONS. Reactants and products are written molecularly, and the equation is balanced from a consideration of change in oxidation number or electron transfer.

Example 1.

$$Pb\underset{-2}{S} + H\overset{+5}{N}O_3 \longrightarrow Pb\underset{+6}{S}O_4 + \overset{+4}{N}O_2 + H_2O$$

$$N: +5 \xrightarrow{\text{①}} +4 \qquad S: -2 \xrightarrow{\text{⑧}} +6$$

1 S gains 8 in oxidation number or loses 8 electrons
1 N loses 1 in oxidation number or gains 1 electron

In balancing, use this ratio 8 to 1; thus,

$$PbS + 8\ HNO_3 \longrightarrow PbSO_4 + 8\ NO_2 + 4\ H_2O$$

Example 2.

$$\overset{0}{Cu} + H\underset{+5}{N}O_3 \longrightarrow \overset{+2}{Cu}(NO_3)_2 + \underset{+2}{N}O + H_2O$$

$$Cu: 0 \xrightarrow{\text{②}} +2 \qquad N: +5 \xrightarrow{\text{③}} +2$$

1 Cu gains 2 in oxidation number or loses 2 electrons
1 N loses 3 in oxidation number or gains 3 electrons

Then 3 Cu will lose 6 electrons; 2 N will gain 6 electrons. Start balancing the equation with the 3 Cu to 2 N ratio

$$3\ Cu + 2\ HNO_3 \rightarrow 3\ Cu(NO_3)_2 + 2\ NO + H_2O$$

Balance the other elements. In this case 6 additional NO_3 radicals are needed, as only part undergoes oxidation-number change in this reaction. Six more molecules of HNO_3 must be used to furnish the 6 NO_3 groups used in formation of 3 $Cu(NO_3)_2$. Hence the final equation will be

$$3\ Cu + 8\ HNO_3 \rightarrow 3\ Cu(NO_3)_2 + 2\ NO + 4\ H_2O$$

Example 3.

$$\overset{0}{Cu} + \text{conc. } H_2\underset{+6}{S}O_4 \xrightarrow{\Delta} \overset{+2}{Cu}SO_4 + \underset{+4}{S}O_2 + H_2O$$

$$Cu: 0 \xrightarrow{\text{②}} +2 \qquad S: +6 \xrightarrow{\text{②}} +4$$

1 Cu gains 2 in oxidation number or loses 2 electrons
1 S loses 2 in oxidation number or gains 2 electrons

Thus, a 1 to 1 ratio of Cu to S is employed in balancing the equation; in this case one additional sulfate radical containing a sulfur unchanged in oxidation number is also needed. The balanced equation is

$$1\ Cu + 2\ H_2SO_4 \xrightarrow{\Delta} CuSO_4 + SO_2 + 2\ H_2O$$

Example 4.

$$Mn(NO_3)_2 + NaBiO_3 + HNO_3 \longrightarrow HMnO_4 + Bi(NO_3)_3, + NaNO_3 + H_2O$$

Mn: 2 → 7 (5); Bi: 5 → 3 (2)

1 Mn gains 5 in oxidation number or loses 5 electrons
1 Bi loses 2 in oxidation number or gains 2 electrons
2 Mn will lose 10 electrons; 5 Bi gain 10 electrons

Start balancing the equation with this Mn to Bi ratio:

$$2\ Mn(NO_3)_2 + 5\ NaBiO_3 + HNO_3 \longrightarrow 2\ HMnO_4 + 5\ Bi(NO_3)_3 + NaNO_3 + H_2O$$

Then balance the other elements or radicals, balancing the water last.

$$2\ Mn(NO_3)_2 + 5\ NaBiO_3 + 16\ HNO_3 \longrightarrow 2\ HMnO_4 + 5\ Bi(NO_3)_3 + 5\ NaNO_3 + 7\ H_2O$$

Example 5. Occasionally an equation is encountered in which more than two elements change in oxidation state. For example:

$$As_2S_3 + HNO_3 + H_2O \rightarrow H_3AsO_4 + NO + S$$

N: +5 → +2 (3); As: +3 → +5 (2); S: −2 → 0 (2)

1 As gains 2 in oxidation number or loses 2 electrons
2 As in As_2S_3 gain 4 in oxidation number
1 S gains 2 in oxidation number or loses 2 electrons
3 S in As_2S_3 gain $3 \times 2 = 6$ in oxidation number
The total change in As_2S_3 is then +4 and $+6 = +10$.

1 N atom in HNO_3 loses 3 in oxidation number or gains 3 electrons.
Since there is only 1 N per molecule of HNO_3, the total change in HNO_3 is −3.
From the above we arrive at the ratio of 3 to 10.
Therefore

$$3\ As_2S_3 + 10\ HNO_3 + 4\ H_2O \rightarrow 6\ H_3AsO_4 + 10\ NO + 9\ S$$

To arrive at the proper coefficients for equations of this type, it is necessary only to determine the total **net** change of oxidation numbers in each molecule of reacting substances.

IONIC EQUATIONS. Substances are shown in ionic form (if ions are present in solution), and the equation may be balanced as above from a consideration of oxidation-number change or electron transfer. Usually only the ions actually undergoing change are shown in the equation.

Example 6.

+5 —— ③ —— +2

$$Cu + H^+ + NO_3^- \rightarrow Cu^{++} + NO + H_2O$$

0 —— ② —— +2

The ratio 3 to 2 is obtained; 3 is placed in front of Cu, and 2 in front of NO_3^-. Other elements are then balanced:

$$3\ Cu + 8\ H^+ + 2\ NO_3^- \rightarrow 3\ Cu^{++} + 2\ NO + 4\ H_2O$$

Example 7.

+7 —— ⑤ —— +2

$$MnO_4^- + AsO_3^{3-} + H^+ \rightarrow Mn^{++} + AsO_4^{3-} + H_2O$$

+3 —— ② —— +5

The balanced equation becomes

$$2\ MnO_4^- + 5\ AsO_3^{3-} + 6\ H^+ \rightarrow 2\ Mn^{++} + 5\ AsO_4^{3-} + 3\ H_2O$$

ION-ELECTRON METHOD. This solution is applicable only to ionic reactions; however, since most of our reactions in solution are between ions, the method proves convenient and useful in writing and balancing equations for such reactions. The method takes into account only those ions in solution which actually enter into reaction; in other words, only the net reaction is shown. Those ions which remain unaltered in the solution do not appear in the equation. The oxidation-reduction equation is broken down into two partial equations or so-called *half-reactions*; one equation represents the oxidation step, the other equation represents the reduction step. The number of electrons transferred in each step is indicated, and each equation is balanced both atomically and electrically. The gain and loss of electrons is then balanced by multiplying each step by the appropriate number. The final single-step equation is obtained by adding the two half-reactions.

Example 8.

$$FeCl_3 + SnCl_2 \rightarrow FeCl_2 + SnCl_4$$

From an ionic standpoint we should write down only the ions which enter into the chemical change: Fe^{3+}, which is reduced to Fe^{++}, and Sn^{++}, which is oxidized to Sn^{4+}. The two half-reactions would be

$$Sn^{++} \rightarrow Sn^{4+} + 2\ e^- \text{ (oxidation)}$$
$$Fe^{3+} + e^- \rightarrow Fe^{++} \text{ (reduction)}$$

Each step is now balanced electrically as well as atomically. Since 2 electrons appear in the first step, we may balance electrons by multiplying the second equation by 2:

$$2\ Fe^{3+} + 2\ e^- \rightarrow 2\ Fe^{++}$$

Now the two steps may be added (electrons are cancelled) and we obtain the net ionic reaction:

$$Sn^{++} \rightarrow Sn^{4+} \cancel{+ 2\ e^-}$$
$$2\ Fe^{3+} \cancel{+ 2\ e^-} \rightarrow 2\ Fe^{++}$$
$$\overline{Sn^{++} + 2\ Fe^{3+} \rightarrow Sn^{4+} + 2\ Fe^{++}}$$

Note that the final step is also balanced electrically; the net charge on each side of the equation is +8.

Example 9. Copper reacts with dilute nitric acid to form cupric ions, water, and nitric oxide. Applying the ion-electron method to this reaction, copper is oxidized from the free state to cupric ion:

$$Cu \rightarrow Cu^{++} + 2\ e^- \text{ (oxidation)} \qquad (1)$$

Meanwhile nitric acid is reduced to NO and water, which are un-ionized:

$$H^+ + NO_3^- \rightarrow NO + H_2O$$

If the latter equation is to balance atomically, 2 molecules of water are necessary to balance the 3 oxygen atoms on the left side of the equation. Four hydrogen atoms are then necessary on the left to balance the 4 hydrogen atoms in the 2 molecules of water, thus:

$$4\ H^+ + NO_3^- \rightarrow NO + 2\ H_2O$$

The equation is balanced atomically but not electrically, since the net charge on the left is +3, and the net charge on the right side is 0. The addition of 3 e^- to the left side will balance the step electrically:

$$3\ e^- + 4\ H^+ + NO_3^- \rightarrow NO + 2\ H_2O \text{ (reduction)} \qquad (2)$$

To balance the gain and loss of electrons, equation (1) may be multiplied by 3 and equation (2) by 2:

$$3\ Cu \rightarrow 3\ Cu^{++} + \cancel{6\ e^-}$$
$$\cancel{6\ e^-} + 8\ H^+ + 2\ NO_3^- \rightarrow 2\ NO + 4\ H_2O$$
$$\overline{3\ Cu + 8\ H^+ + 2\ NO_3^- \rightarrow 3\ Cu^{++} + 2\ NO + 4\ H_2O}$$

Example 10. It is further assumed in the ion-electron method that H^+, OH^-, and H_2O are available for balancing each step of the equation atomically. To illustrate, permanganate ion will oxidize sulfide ion to free sulfur in acid solution

$$S^{--} \rightarrow S + 2\ e^- \qquad (3)$$

and the permanganate ion is reduced to manganous ion

$$MnO_4^- \rightarrow Mn^{++}$$

To balance the latter step atomically, we may add 8 hydrogen ions to the left side, in which case 4 molecules of water will appear on the right side:

$$8\ H^+ + MnO_4^- \rightarrow Mn^{++} + 4\ H_2O$$

Five electrons are then necessary on the left side to balance electrically:

$$5\ e^- + 8\ H^+ + MnO_4^- \rightarrow Mn^{++} + 4\ H_2O \qquad (4)$$

Combining steps (3) and (4) and balancing gain and loss of electrons,

$$5\ S^{--} \rightarrow 5\ S + \cancel{10\ e^-}$$
$$\cancel{10\ e^-} + 16\ H^+ + 2\ MnO_4^- \rightarrow 2\ Mn^{++} + 8\ H_2O$$
$$\overline{5\ S^{--} + 16\ H^+ + 2\ MnO_4^- \rightarrow 5\ S + 2\ Mn^{++} + 8\ H_2O}$$

Oxidation-reduction potentials. In using the ion-electron half-reaction method for representing oxidation-reduction processes, electrons are shown as being given up by the reducing agent and taken up by the oxidizing agent. The tendency of an oxidizing agent to take up electrons and a reducing agent to receive electrons may be measured and expressed as a potential or voltage (see text). An abbreviated table of voltages (E^0) is shown below. In this table all of the half-reactions are written as oxidation processes with electrons on the right side of the equation, and the voltages recorded are termed *oxidation* potentials. By reversing the equation and also the sign of the voltage, the corresponding *reduction* potentials are obtained.

TABLE 4. *An Abbreviated Table of Oxidation-Reduction Potentials*

Half-reaction			E^0 *
K	⇆	$K^+ + e^-$	2.92
Na	⇆	$Na^+ + e^-$	2.71
Al	⇆	$Al^{3+} + 3\ e^-$	1.66
Mn	⇆	$Mn^{++} + 2\ e^-$	1.18
AsH_3	⇆	$As + 3\ H^+ + 3\ e^-$	0.60
Cr^{++}	⇆	$Cr^{3+} + e^-$	0.41
$Pb + SO_4^{--}$	⇆	$PbSO_4 + 2\ e^-$	0.36
Ni	⇆	$Ni^{++} + 2\ e^-$	0.25
Sn	⇆	$Sn^{++} + 2\ e^-$	0.14
H_2	⇆	$2\ H^+ + 2\ e^-$	0.00
PH_3	⇆	$P + 3\ H^+ + 3\ e^-$	−0.06
$2\ S_2O_3^{--}$	⇆	$S_4O_6^{--} + 2\ e^-$	−0.08
H_2S	⇆	$S + 2\ H^+ + 2\ e^-$	−0.14
$H_2O + H_2SO_3$	⇆	$SO_4^{--} + 4\ H^+ + 2\ e^-$	−0.20
Cu	⇆	$Cu^{++} + 2\ e^-$	−0.34
$S + 3\ H_2O$	⇆	$H_2SO_3 + 4\ H^+ + 4\ e^-$	−0.45
$2\ I^-$	⇆	$I_2 + 2\ e^-$	−0.53
$H_3AsO_3 + H_2O$	⇆	$H_3AsO_4 + 2\ H^+ + 2\ e^-$	−0.56
H_2O_2	⇆	$O_2 + 2\ H^+ + 2\ e^-$	−0.68
Fe^{++}	⇆	$Fe^{3+} + e^-$	−0.77
Ag	⇆	$Ag^+ + e^-$	−0.80
$H_2O + NO_2$	⇆	$NO_3^- + 2\ H^+ + e^-$	−0.81
$NO + H_2O$	⇆	$HNO_2 + H^+ + e^-$	−0.99
$2\ Br^-$	⇆	$Br_2 + 2\ e^-$	−1.06
$2\ H_2O + Mn^{++}$	⇆	$MnO_2 + 4\ H^+ + 2\ e^-$	−1.28
$2\ Cr^{3+} + 7\ H_2O$	⇆	$Cr_2O_7^{--} + 14\ H^+ + 6\ e^-$	−1.33
$2\ Cl^-$	⇆	$Cl_2 + 2\ e^-$	−1.36
$Pb^{++} + 2\ H_2O$	⇆	$PbO_2 + 4\ H^+ + 2\ e^-$	−1.46
$Cl_2 + 2\ H_2O$	⇆	$2\ HClO + 2\ H^+ + 2\ e^-$	−1.63
Au	⇆	$Au^+ + e^-$	−1.68
$MnO_2 + 2\ H_2O$	⇆	$MnO_4^- + 4\ H^+ + 3\ e^-$	−1.70
Co^{++}	⇆	$Co^{3+} + e^-$	−1.84
$O_2 + H_2O$	⇆	$O_3 + 2\ H^+ + 2\ e^-$	−2.07
$2\ F^-$	⇆	$F_2 + 2\ e^-$	−2.85

* E^0 values are for the following conditions: 1 M aqueous solution, 25°C, 1 atm.

Such a table as above enables one to predict whether or not a given oxidation-reduction reaction will occur. Any combination of an oxidation half-reaction and a reduction half-reaction may be added together along with the potentials of each half-reaction. If the sum of the voltages of half-reactions is positive, then the overall single-step equation as obtained from adding the two half-reactions will proceed spontaneously. Conversely if the voltage sum is negative the reaction does not proceed. Study the examples below which illustrate the use of oxidation-reduction potentials.

Example 1. Consider the two half-reactions

$$Al \rightarrow Al^{3+} + 3\,e^- \qquad E^0 = 1.66 \text{ (oxidation)} \qquad (1)$$

$$Cu \rightarrow Cu^{++} + 2\,e^- \qquad E^0 = -0.34 \text{ (oxidation)} \qquad (2)$$

By reversing either half-reaction we obtain a reduction half-reaction; for example, if we reverse (2), then

$$Cu^{++} + 2\,e^- \rightarrow Cu \qquad E^0 = +0.34 \text{ (reduction)} \qquad (3)$$

Add (1) and (3) (after multiplying to balance electrons):

$$\begin{array}{ll} 2\,Al \rightarrow 2\,Al^{3+} + 6\,e^- & E^0 = 1.66 \\ 6\,e^- + 3\,Cu^{++} \rightarrow 3\,Cu & E^0 = 0.34 \\ \hline 2\,Al + 3\,Cu^{++} \rightarrow 2\,Al^{3+} + 3\,Cu & E\ = 2.00 \end{array}$$

Since the voltage is positive the reaction will occur as written. If we had chosen to write (1) as a reduction process, that is, $Al^{3+} + 3\,e^- \rightarrow Al$ ($E^0 = -1.66$), and added this to (2) the total voltage would be -2.00 and the reaction $2\,Al^{3+} + 3\,Cu \rightarrow 2\,Al + 3\,Cu^{++}$ would not occur. Note in the above that the values of E^0 are not changed when the half-reactions are multiplied to effect a balance of electrons.

Example 2. Will NO_3^- oxidize H_2SO_3 in acid solution?

Consider the two half-reactions from the table:

$$\begin{array}{ll} H_2SO_3 + H_2O \rightarrow SO_4^{--} + \cancel{4\,H^+} + \cancel{2\,e^-} & E^0 = -0.20 \\ \cancel{2\,e^-} + 2\,NO_3^- + \cancel{4\,H^+} \rightarrow 2\,H_2O + 2\,NO_2 & E^0 = \ \ 0.81 \\ \hline H_2SO_3 + 2\,NO_3^- \rightarrow SO_4^{--} + 2\,NO_2 + H_2O & E\ = \ \ 0.61 \end{array}$$

Since E for the cell is positive, the reaction proceeds as written.

Example 3. Consider combination of the two half-reactions

$$\begin{array}{ll} 6\,F^- \rightarrow 3\,F_2 + 6\,e^- & E^0 = -2.85 \\ 6\,e^- + Cr_2O_7^{--} + 14\,H^+ \rightarrow 2\,Cr^{3+} + 7\,H_2O & E^0 = \ \ 1.33 \\ \hline 6\,F^- + Cr_2O_7^{--} + 14\,H^+ \rightarrow 3\,F_2 + 2\,Cr^{3+} + 7\,H_2O & E\ = -1.52 \end{array}$$

Since E has a negative value this reaction will not proceed.

Formation of double salts. A number of naturally occurring crystalline minerals have formulas that correspond to a combination of two salts—for example, the minerals *carnallite* ($KCl \cdot MgCl_2 \cdot 6H_2O$) and *schönite* ($K_2SO_4 \cdot MgSO_4 \cdot 6H_2O$). As a solution of mixed salts crystallizes, a fixed-composition crystalline compound is frequently formed between them. Some commonly prepared compounds of this nature are Mohr's salt [$(NH_4)_2SO_4 \cdot FeSO_4 \cdot 6H_2O$] and an alum [$Na_2SO_4 \cdot Al_2(SO_4)_3 \cdot 24H_2O$]. Salts of this kind, which in solution give the individual ions of the component salts, are called *double salts.* Such salts in solution exhibit chemical changes appropriate for each respective ion.

Formation of complex salts and complex ions. Certain crystalline compounds that may by composition appear to be double salts do not conform to the above definition of a double salt. A formula for a compound that could be written $4KCN \cdot Fe(CN)_2$ is usually written $K_4Fe(CN)_6$. There is definite experimental evidence to prove that little Fe^{++} is present in its water solution. It ionizes as follows:

$$K_4Fe(CN)_6 \rightarrow 4\ K^+ + Fe(CN)_6^{4-}$$

Other examples are:

$$PbF_2 \cdot SiF_4 \quad \text{or} \quad PbSiF_6 \rightarrow Pb^{++} + SiF_6^{--}$$

$$3NaF \cdot AlF_3 \quad \text{or} \quad Na_3AlF_6 \rightarrow 3\ Na^+ + AlF_6^{3-}$$

$$2NaCl \cdot PtCl_4 \quad \text{or} \quad Na_2PtCl_6 \rightarrow 2\ Na^+ + PtCl_6^{--}$$

$$3Na_2S \cdot As_2S_5 \quad \text{or} \quad 2\ Na_3AsS_4 \rightarrow 6\ Na^+ + 2\ AsS_4^{3-}$$

These compounds are called *complex salts* since, in their ionization, only two ions are formed, a single element ion and a *complex ion.*

Complex ions are formed by combination of an ion with other ions or neutral molecules. Ordinarily the units in such a complex ion are held together by coordinate bonds of valence—that is, a pair of electrons furnished by one of the ions or neutral molecules. The hydronium ion, H_3O^+, and the ammonium ion, NH_4^+, may be considered complex ions, since a neutral molecule of water or ammonia furnishes a pair of electrons to combine with a proton.

Common examples of complex ion formation are those in which a metallic ion, such as Cu^{++}, Co^{3+}, or Ag^+, combines with a molecule or another ion to form a complex. Ammonia forms complex ions with a number of cations, for example,

$$Cu^{++} + 4\ NH_3 \rightarrow Cu(NH_3)_4^{++}$$

$$Co^{3+} + 6\ NH_3 \rightarrow Co(NH_3)_6^{3+}$$

$$Ag^+ + 2\ NH_3 \rightarrow Ag(NH_3)_2^+$$

$$H^+ + NH_3 \rightarrow NH_4^+$$

Similarly, complex anions may be formed:

$$S^{--} + S \rightarrow S_2^{--} \text{ (in polysulfides)}$$

In each of the above cases the bond of the neutral molecule to the ion is one in which both electrons of the bond are contributed by the neutral molecule. This can best be shown in the following diagrams, in which the electrons contributed by N and S are shown as dots, those by H as crosses; Cu^{++} has no valence electrons.

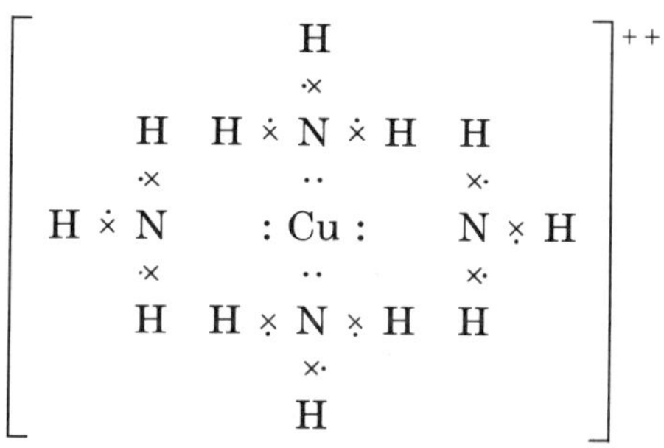

$Cu(NH_3)_4^{++}$

S_2^{--}

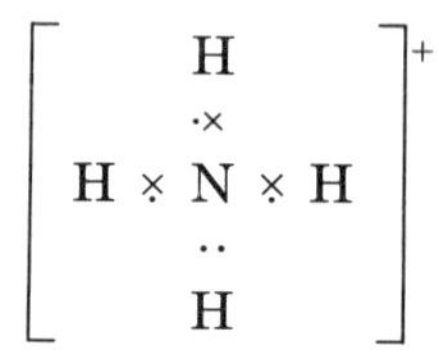

NH_4^+

If NH_4OH is added to a solution of Cu^{++}, $Cu(OH)_2$ is first precipitated, but it redissolves as more NH_4OH is added because of the formation of $Cu(NH_3)_4^{++}$. The latter ion has a deep blue color. $Cu(OH)_2$, which is insoluble in water, is therefore readily soluble in an aqueous solution containing free ammonia. Other hydroxides are soluble in NH_4OH if the cation forms a complex with NH_3. Consequently, the formation of complex ions may be important in solubility considerations. In many cases, the number of molecules or ions combined with a simple ion is double the valence of the simple cation. Thus Cu^{++} usually coordinates four such groups, Ag^+ two, Co^{3+} six, and so on. The number of molecules or ions grouped about a central ion is referred to as the *coordination number.*

In qualitative analysis, complex sulfide salts, or ions thereof, are encountered in various separations. HgS will dissolve in a Na_2S solution, whereas PbS will not, because HgS forms a complex ion.

$$HgS + Na_2S \rightarrow Na_2S \cdot HgS \quad \text{or} \quad Na_2HgS_2$$

or

$$HgS + S^{--} \rightarrow HgS_2^{--}$$

Similarly,

$$As_2S_5 + 3\ Na_2S \rightarrow 3Na_2S \cdot As_2S_5 \quad \text{or} \quad 2\ Na_3AsS_4$$

or

$$As_2S_5 + 3\ S^{--} \rightarrow 2\ AsS_4^{3-}$$

Cyanide ion, CN^-, readily forms complexes with several metallic ions, for example,

$$Ag^+ + 2\ CN^- \rightarrow Ag(CN)_2^-$$

$$Fe^{3+} + 6\ CN^- \rightarrow Fe(CN)_6^{3-}$$

Silver cyanide, AgCN, which is insoluble in water, is readily soluble in excess CN^- because of the formation of the complex $Ag(CN)_2^-$.

Halide ions frequently act similarly to form complex ions, for example, $AuCl_4^-$, $CuCl_4^{--}$, SiF_6^{--}, $SbCl_6^{3-}$, $PtCl_6^{--}$, and AlF_6^{3-}.

The structures of the $Co(NH_3)_6^{3+}$, $AuCl_4^-$, and AsS_4^{3-} ions are shown below. $Co(NH_3)_6^{3+}$

has an octahedral structure, $AuCl_4^-$ is square planar, and AsS_4^{3-} is probably tetrahedral. Other complexes have somewhat similar structures.

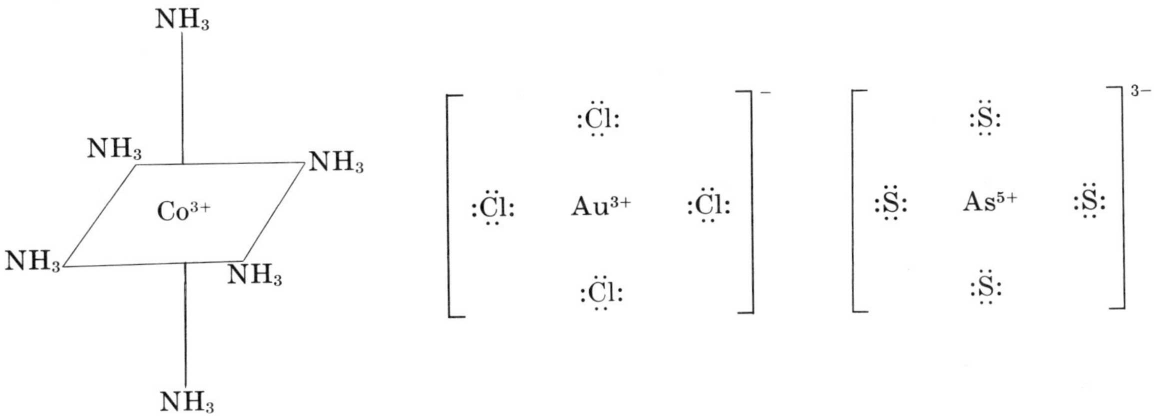

(octahedral) (square planar) (tetrahedral)

The charge of the complex ions can be determined by considering the charge of the central cation and then the number of extra electrons added. The $Co(NH_3)_6^{3+}$ ion retains the charge of the Co^{3+}, since the 6 added ammonia molecules are neutral. The $AuCl_4^-$ has its mononegative charge, since the Au^{3+} ion has had 4 Cl^- ions added. The AsS_4^{3-} ion can be construed as a As^{5+} ion with 4 S^{--} ions added.

Theory of complex ion formation. In the examples above, it may be noted that a positive ion (cation) is attached or combined with other ions (anions) or neutral molecules. The latter have unused pairs of electrons which are donated to the cation to form coordinate bonds. The positive ion may be regarded as an electron pair *acceptor*.

The number of electron pairs accepted (coordinate bonds formed) by a cation depends on its sub-orbital structure. For example, the $Zn(NH_3)_4^{++}$ complex ion stability is related to orbital structures as follows:

$_{30}Zn^{65}$ has the electronic configuration $1s^2$; $2s^2$, $2p^6$; $3s^2$, $3p^6$, $3d^{10}$; $4s^2$.

The outer structure may be represented as

Zn^{++} ion may be represented as

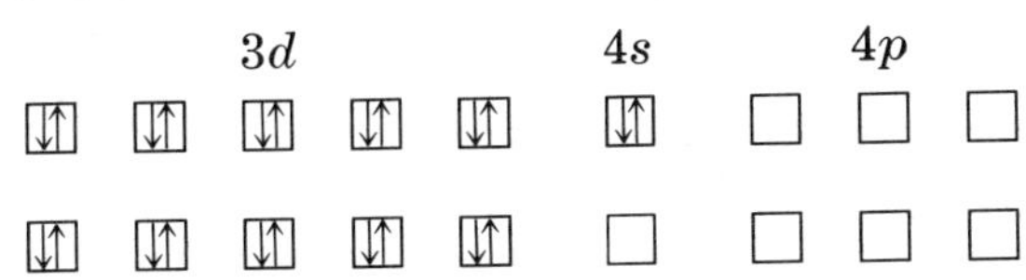

Each of 4 NH_3 molecules, $H{:}\ddot{N}{:}H$ (with $\ddot{H}$ below N), has an unused or unshared pair to donate to the vacant one $4s$ and the three $4p$ orbitals. Consequently, 4 molecules of NH_3 attach to the Zn^{++} to give it a coordination number of 4.

Another example is in formation of the complex ion, $Fe(CN)_6^{4-}$:

$_{26}Fe^{56}$ atom has the configuration $1s^2$; $2s^2$, $2p^6$; $3s^2$, $3p^6$, $3d^6$; $4s^2$.

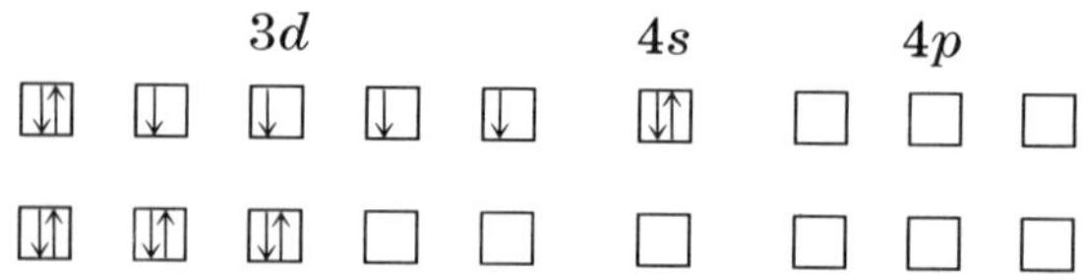

Fe^{++} ion may be represented as

Each of 6 CN^- (cyanide), :C:::N:, ions donates an electron pair to complete the 3*d*, 4*s*, and 4*p* orbitals; thus Fe^{++} has a coordination number of 6.

Occasions arise when a given orbital will have only 1 electron; for example, the complex ion $Cu(CN)_4^{--}$ may be represented

Cu^{++}

$Cu(CN)_4^{--}$

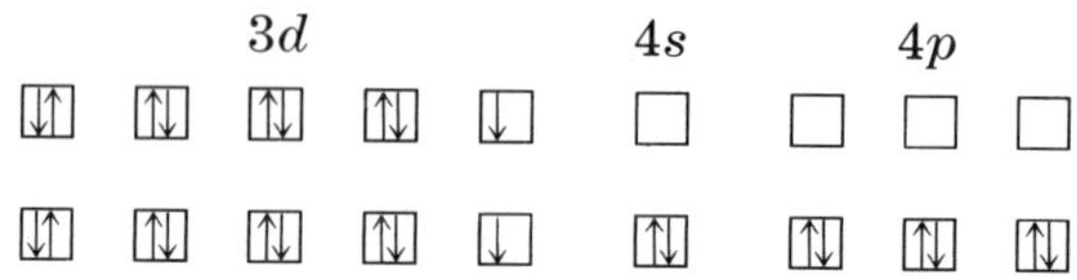

The coordination number of Cu^{++} in this case is 4, as also in $Cu(NH_3)_4^{++}$.

Ag^+ and Au^+ have usual coordination numbers of 2; a less common coordination number of 8 is apparent in some compounds of molybdenum.

The tendency to form coordination compounds is much greater with transition metal ions than with alkali or alkaline earth ions. The latter have large ionic diameters, small charge, and no incomplete *d* orbitals—factors which undoubtedly detract from their ability to form complex ions.

Coordinating groups are called *ligands* and may be either ions or neutral molecules. Almost all anions may act as electron pair donors. Common molecules which act as ligands are H_2O, NH_3, N_2H_4, and C_2H_5OH.

Amphoterism. Strong metallic hydroxides, such as NaOH, are basic. The hydroxides of nonmetals such as $P(OH)_3$ are acidic and, consequently, are usually not written in the same manner as the bases. For example, phosphorous acid is written H_3PO_3 instead of $P(OH)_3$. There are many slightly soluble metallic hydroxides, however, which may act either as an acid or as a base, and, accordingly, may be dissolved by either a base or an acid. For example, $Al(OH)_3$ is soluble either in hydrochloric acid or sodium hydroxide and is said to be amphoteric.

$$Al(OH)_3 + 3\ H^+ \rightarrow Al^{3+} + 3\ H_2O$$

$$(H_3AlO_3) \text{ or } Al(OH)_3 + 3\ OH^- \rightarrow Al(OH)_6^{3-} \text{ or } AlO_3^{3-}\bullet 3H_2O$$

and $$(H_3AlO_3) \text{ or } Al(OH)_3 + OH^- \rightarrow Al(OH)_4^- \rightarrow AlO_2^-\bullet 2H_2O$$

The solubility in the strong base is due to the formation of a complex ion, in this case the ortho-aluminate ion, AlO_3^{3-}, or meta-aluminate ion, AlO_2^- (omitting water of hydration).

Various qualitative separations are based on the property of amphoterism. If excess NaOH is added to a mixture of ferric and aluminum salts, the ferric hydroxide remains insoluble, and the aluminum hydroxide first precipitated redissolves in the excess base to give a soluble aluminate.

Other amphoteric hydroxides are $Cr(OH)_3$, $Zn(OH)_2$, and $Sn(OH)_2$. They will act with either H^+ or OH^-.

Hydrolysis. Although water is only slightly ionized according to the equation

$$H_2O + H_2O \rightleftharpoons H_3O^+ + OH^-$$

or

$$H_2O \rightleftharpoons H^+ + OH^- \quad (1)$$

the few H^+ ions and OH^- ions produced may become of primary importance in aqueous solutions of certain salts. As long as the number of H^+ ions is equal to the number of OH^- ions the solution is neutral. However, if something is added which furnishes either of the ions or which uses up either of the ions, then 1 of the 2 ions is present in excess. If excess H^+ ions are present, the solution is acid; if excess OH^- ions are present, the solution is basic.

Ions of salts may disturb the equilibrium (1) in aqueous solution, and as a result the solution is actually acidic or basic, depending on whether H^+ or OH^- ions are present in excess. This reaction of the ions of a salt with water is termed *hydrolysis*. We may illustrate hydrolysis with a few examples:

1. SODIUM CARBONATE. An aqueous solution of this salt tests basic toward litmus, indicating an excess of OH^- ions present in the solution. This may readily be explained as follows: Sodium carbonate, a salt derived from the strong base sodium hydroxide and the weak acid carbonic acid, is completely ionized into Na^+ and CO_3^{--}. CO_3^{--} ions have a great affinity for protons to form the weak (slightly ionized) acid HCO_3^-; consequently, protons are removed from solution by the reaction

$$H^+ + CO_3^{--} \rightarrow HCO_3^-$$

As a result, the equilibrium (1) is shifted in the direction of a higher concentration of OH^- ions, which gives the solution a basic reaction. Meanwhile there has been no tendency for Na^+ and OH^- ions to combine, since NaOH is a strong base.

2. AMMONIUM CHLORIDE. The ions present in the solution are H^+, OH^-, NH_4^+, and Cl^-. There is very little tendency for H^+ to combine with Cl^-, since HCl is a strong acid. However, NH_4^+ does combine with OH^-, since NH_4OH (or $NH_3 + H_2O$) is a weak base. Accordingly, an excess of H^+ ions results, and the solution reacts acid to litmus.

3. SODIUM CHLORIDE. Ions present in the solution are Na^+, Cl^-, H^+, and OH^-. Since both NaOH and HCl are strong electrolytes, there is little if any tendency for either to be formed. Hence the concentrations H^+ and OH^- ions remain about equal, and the resulting solution is approximately neutral. The hydrolysis of a salt derived from a strong acid and strong base is negligible.

4. AMMONIUM ACETATE. The aqueous solution contains NH_4^+, $C_2H_3O_2^-$, H^+, and OH^- ions. This salt is hydrolyzed, since both acetic acid and ammonium hydroxide are weak electrolytes:

$$NH_4^+ + C_2H_3O_2^- + H_2O \rightleftharpoons NH_4OH + HC_2H_3O_2$$

Ammonium hydroxide and acetic acid are of about the same strength as electrolytes; consequently, about the same number of H^+ and OH^- ions are left in the solution, which is thus approximately neutral.

5. COPPER SULFATE. Salts of most of the heavy metals contain hydrated ions which may remove OH^- from the solution and give it an acidic reaction. For example, copper sulfate in solution gives $Cu(H_2O)_4^{++}$ and SO_4^{--} ions. There would be little tendency for SO_4^{--} to combine with H^+, since H_2SO_4 is a strong acid, but the hydrated cupric ions may react as follows:

$$Cu(H_2O)_4^{++} + OH^- \rightleftharpoons Cu(H_2O)_3OH^+ + H_2O$$

Thus a solution of this salt is slightly acidic because, as OH^- ions are removed and more water molecules ionize, an increase of H^+ ion concentration over that of pure water develops.

Hydrolysis constants. Since hydrolysis is an equilibrium process, an equilibrium constant may be determined. The value of this constant will be a measure of the degree or extent of the hydrolysis process. Consider the hydrolysis of $NaC_2H_3O_2$:

$$Na^+ + C_2H_3O_2^- + H_2O \rightleftharpoons Na^+ + HC_2H_3O_2 + OH^-$$

$$K_i = \frac{[HC_2H_3O_2][OH^-]}{[C_2H_3O_2^-][H_2O]} \qquad (1)$$

Since $[H_2O]$ is about constant in dilute solutions, we may incorporate its constant value into K_i and obtain an expression for K_h, called the hydrolysis constant:

$$K_i \times [H_2O] = K_h = \frac{[HC_2H_3O_2][OH^-]}{[C_2H_3O_2^-]} \qquad (2)$$

The larger the value of K_h, the greater the degree or extent of hydrolysis, that is, the farther the equilibrium is displaced to the right side of the equation as written. K_h has a value of 5.5×10^{-10} for the hydrolysis of $C_2H_3O_2^-$. The small number indicates that the reaction as written does not proceed very far to the right; that is, the hydrolysis is not high.

We have previously learned that a weak acid is always in equilibrium with its ions. The relationship is shown by the ionization constant for that acid. For example, in (2) above, acetic acid is in equilibrium with H^+ and $C_2H_3O_2^-$:

$$HC_2H_3O_2 \rightleftharpoons H^+ + C_2H_3O_2^-$$

and

$$K_a = \frac{[H^+][C_2H_3O_2^-]}{[HC^2H_3O_2]} = 1.8 \times 10^{-5} \qquad (3)$$

where K_a is the ionization constant for the weak acid.

Also, H_2O is always in equilibrium with H^+ and OH^-,

$$H_2O \rightleftharpoons H^+ + OH^-$$

and the relationship is given by the expression

$$K_w = [H^+][OH^-] = 1 \times 10^{-14} \tag{4}$$

If we now solve equation (4) for $[OH^-]$,

$$[OH^-] = \frac{K_w}{[H^+]}$$

and substitute this value for $[OH^-]$ into (2), we obtain

$$K_h = \frac{[HC_2H_3O_2]\,K_w}{[C_2H_3O_2^-][H^+]} \tag{5}$$

We note next that $\dfrac{[HC_2H_3O_2]}{[C_2H_3O_2^-][H^+]} = \dfrac{1}{K_a}$ from (3) and, substituting in (5), we obtain

$$K_h = \frac{K_w}{K_a}$$

Since $\qquad K_w = 1 \times 10^{-14}$ and $K_a = 1.8 \times 10^{-5}$

$$K_h\ (\text{for } C_2H_3O_2^-) = \frac{1 \times 10^{-14}}{1.8 \times 10^{-5}} = 5.5 \times 10^{-10}$$

It is evident that we may calculate hydrolysis constants for anions of weak acids from the values of K_w and K_a for the weak acid. Similarly, we can arrive at the expression

$$K_h = \frac{K_w}{K_b}$$

for the hydrolysis of a cation of a weak base where K_b is the ionization constant for the weak base. If the hydrolysis is for a salt of both a weak acid and a weak base,

$$K_h = \frac{K_w}{K_a K_b}$$

(The student should verify the last two expressions for K_h. Follow the same procedure as above for deriving the expression for the hydrolysis of $C_2H_3O_2^-$.) Study the following examples:

Problem 1. Calculate K_h for NH_4Cl, given K_b for $NH_4OH = 1.8 \times 10^{-5}$.

$$NH_4^+ + \cancel{Cl^-} + H_2O \rightleftharpoons NH_4OH + H^+ + \cancel{Cl^-}$$

$$K_h = \frac{[NH_4OH][H^+]}{[NH_4^+]} = \frac{K_w}{K_b} = \frac{1 \times 10^{-14}}{1 \times 10^{-8}} = 1 \times 10^{-6}$$

Problem 2. Calculate the approximate pH of 0.1 M NaCN, given K_a for HCN $= 5 \times 10^{-10}$.

$$\cancel{Na^+} + CN^- + H_2O \rightleftharpoons HCN + OH^- + \cancel{Na^+}$$

$$K_h = \frac{[HCN][OH^-]}{[CN^-]} = \frac{K_w}{K_a} = \frac{1 \times 10^{-14}}{5 \times 10^{-10}} = 2 \times 10^{-5}$$

Since $$[OH^-] = [HCN]$$

then $$K_h = \frac{[OH^-]^2}{0.1} = 2 \times 10^{-5}$$

$$[OH^-] = \sqrt{2 \times 10^{-6}} = 1.4 \times 10^{-3}$$

$$pOH = -\log [1.4 \times 10^{-3}] = 2.85 \quad \text{and} \quad pH = 14 - 2.85 = 11.15$$

Part two

The semimicro plan of qualitative analysis

11
The semimicro method

Macro methods of qualitative analysis involve the use of rather large quantities of materials; in the case of solids, amounts of 1 g or more are often used; in the case of liquids, volumes of 100 ml or more are common. The semimicro method is an adaptation of the macro method of qualitative analysis to very much smaller quantities of substances—that is, milligram amounts of solids and drop or milliliter quantities of liquids.

Although tests in semimicro procedures are not more sensitive than those of the macro methods, the former have a great advantage in the saving of time and chemicals. Whereas filtrations, washing of precipitates, and evaporations may require up to an hour on the macro scale, these operations can be carried out in a few minutes on the semimicro basis. As a result, the student can analyze more samples in the time available and thus gain a much wider training in analytical procedures.

Terms used in qualitative analysis.

Centrifuge (cfg.) is a device used to accelerate the settling of precipitates by taking advantage of centrifugal force.

Colloidal precipitates are insoluble particles which have a size between those in suspension and those in true solution. They show Brownian movement and Tyndall effect, and do not readily settle from solution.

Decantation is pouring liquid from a settled precipitate.

Filtration is the separation of a precipitate from a liquid by passage through a porous medium. The *filtrate* is the solution passing through the filter. The *residue* or precipitate remains on the filter paper. (In semimicro analysis, separations are usually performed by centrifuging and decantation.)

Ignition is the strong heating of solids.

Precipitation (pptn.) is the formation of an insoluble solid by the reaction of a reagent on a solution. The insoluble solid is called the *precipitate* (ppt.)

Reagent is a substance used to produce a chemical change in the material under consideration. For example, HCl is a reagent when added to a solution of $AgNO_3$ because it produces a precipitate of AgCl.

Turbidity is a cloudy condition due to suspended matter in a solvent.

Washing is the cleansing of a precipitate (usually by H_2O) to remove the adhering solution from which it has been precipitated.

Semimicro equipment. In general, items of semimicro equipment are much smaller and simpler than the conventional pieces of macro equipment. (A recommended list of items of equipment is given in the Appendix.) One of the features of the semimicro method is the use of a centrifuge (Figure 1) in separating precipitates from solutions. Centrifuges, which should be conveniently located near the working spaces in the laboratory, may be shared by several students.

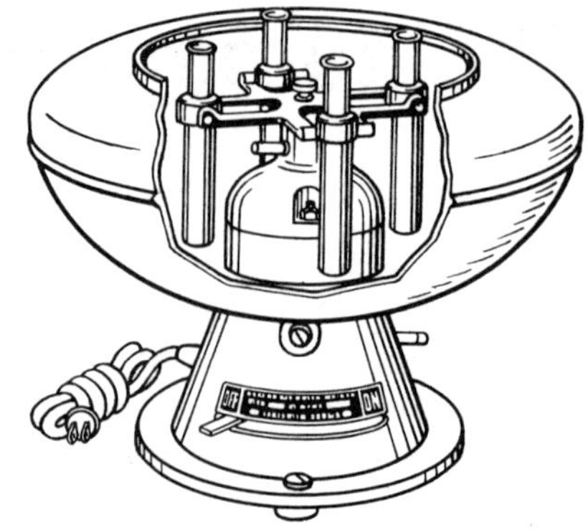

FIGURE 1

There are various ways in which chemicals may be provided and arranged for semimicro work. Each laboratory bench may be equipped with a set of chemicals to be shared by several students. Each student may have a small set of frequently used chemicals, such as the common acids and bases. Several students could then share a larger set of less commonly used reagents. In general, liquids and solutions should be made available in dropping bottles of 15 to 100 ml capacity. Solids may be stored in wide-mouth salt bottles. An arrangement which has been used successfully is described below.

Each student outfit should include a block of liquid reagents which are used frequently—for example, dil. and conc. HCl, dil. and conc. H_2SO_4, and so on. The dropping bottles may be filled from side shelf bottles as the need arises. These individual re-

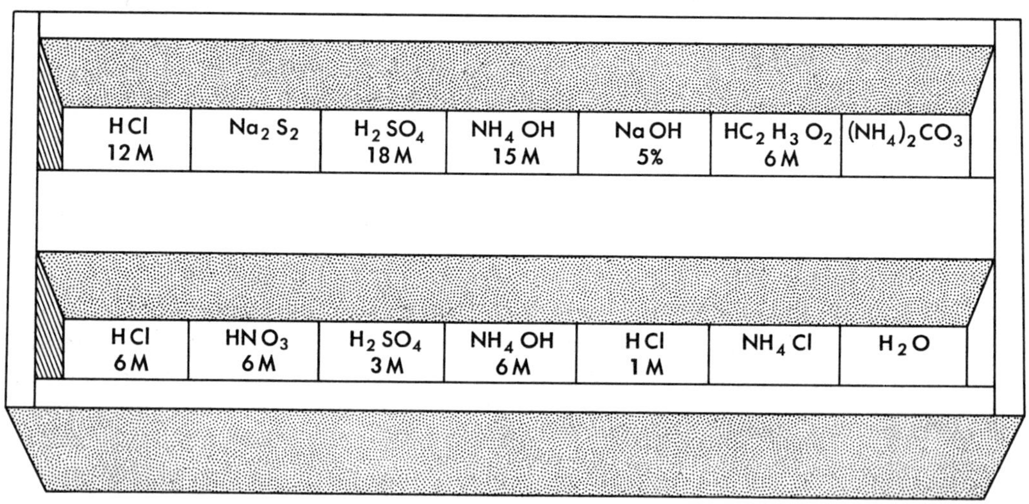

FIGURE 2

agent sets consist of 14 square dropping bottles with screw caps (15 ml), which are housed in a grooved block of wood. The latter is easily constructed by making two parallel grooves (3/4″ deep) in a piece 2″ × 4″ dimension lumber (Figure 2). The ends may be closed with strips of plywood. Less frequently used liquid and solid reagents are stored in larger bottles (about 100 ml) placed on side shelves. Bottles for liquid reagents should be equipped with medicine droppers and bulbs; bottles of solids may be equipped with spatulas inserted through the stoppers. A convenient test tube rack for 3″ or 4″ test tubes may be constructed from a $6\frac{1}{2}''$ piece of $2'' \times 4''$ dimension lumber by drilling holes (17) about $\frac{1}{2}''$ in diameter and 1″ deep.

Laboratory techniques. Probably the most important technique to develop in semimicro analysis is the proper handling of precipitates. Precipitations are usually carried out in small (3″ or 4″) test tubes by the drop-by-drop addition of reagents from medicine droppers. The tube is then centrifuged to settle the precipitate. Complete precipitation is essential and should be tested for by the addition of a drop of the precipitating reagent to the supernatant liquid above the precipitate after centrifuging. If no further precipitation takes place in the supernatant liquid, precipitation may be assumed complete.

The centrifuge must be balanced by placing a tube of the same size containing an amount of water equal to the weight of the test solution opposite to the tube containing the test solution. If the centrifuge vibrates, it should be turned off and the trouble investigated. The time of centrifuging depends on the nature of the precipitate, but 15–20 seconds will usually suffice. As a result of the centrifuging, the solid becomes packed tightly in the end of the tube. The supernatant liquid (sometimes referred to as *decantate*) may be removed by decanting without disturbing the precipitate, or may be drawn off by inserting a finely drawn tube fitted with a medicine dropper bulb to a point just above the precipitate in the tube; then by releasing the pressure on the bulb, the supernatant liquid is drawn up into the tube.

After the precipitate and solution are separated, the precipitate is washed by adding a few drops of the washing reagent (usually water), mixing thoroughly with a stirring rod, centrifuging, and removing the washings with the aid of a medicine dropper tube or by decantation. Two or more washings are usually necessary. Failure to wash precipitates thoroughly is one of the most common sources of error in qualitative analysis.

Evaporations are usually carried out in a small evaporating dish over a beaker of boiling water, which serves as a hot water bath, or by cautious heating of a solution in a crucible with a low flame to prevent spattering.

Absolute cleanliness is essential to success in semimicro analysis. Great care should be taken to prevent contamination of liquid or solid reagents. It is good practice to cover the working space on the laboratory table with a clean towel. A beaker of clean water reserved for clean droppers, stirring rods, and spatulas is also helpful.

Solubility rules.* Since solubility of substances in water is an important consideration in analysis, the student should be familiar with the following solubility rules.

* These brief solubility rules are not all-inclusive; the consideration of less common compounds would call for marked augmentation of the rules. For example, this first solubility rule needs additional notes such as: thallous, aurous, and cuprous halides are insoluble. Salts such as $SnCl_4$, $BiCl_3$, or $SbCl_3$ hydrolyze so markedly when put in water that acid is needed to keep them in solution, for example,

$$BiCl_3 + H_2O \rightleftharpoons BiOCl\downarrow + 2\ HCl$$

1. All chlorides, bromides, and iodides are soluble in water, except those of lead, silver, and univalent mercury; mercuric iodide is also insoluble in water. Lead chloride is quite soluble in hot water.
2. All nitrates, nitrites, chlorates, and acetates are soluble in water.
3. All potassium, sodium, and ammonium salts are soluble in water, except K_2PtCl_6, $(NH_4)_2PtCl_6$, $KHC_4H_4O_6$, $K_3Co(NO_2)_6$, Na_2SiF_6, and $NaSb(OH)_6$, which are sparingly soluble.
4. All oxides and hydroxides are insoluble in water except those of the alkali metals and those of barium, strontium, and calcium, which are sparingly soluble.
5. All sulfides except those of the alkali and alkaline earth metals are insoluble in water.
6. The sulfates of barium, strontium, and lead are insoluble in water and acids. The sulfate of calcium is slightly soluble. All other sulfates are soluble.
7. All carbonates, phosphates, borates, oxalates, chromates, arsenites, arsenates, ferrocyanides, and ferricyanides except those of the alkali metals are insoluble or only very slightly soluble in water ($CaCrO_4$ and $MgCrO_4$ are fairly soluble). All are readily soluble in dilute reagent acids.
8. All silicates are insoluble in water except those of the alkali metals.
9. All cyanides are insoluble in water except those of the alkali metals and bivalent mercury.

12

Schedule of laboratory work

A suggested schedule of laboratory work for a course in semimicro qualitative analysis is outlined below.

A. Preparation of special equipment
B. Preliminary experiment on laboratory techniques
C. Tests for anions
   1. Sulfuric acid treatment
   2. Group tests
   3. Specific tests
D. Analysis of anion unknowns
E. Preliminary experiment on group separations
F. Analysis for cations by groups
   1. Group I. The silver group
      a) Preliminary tests
      b) Analysis of known solution
      c) Analysis of unknown solution
   2. Group II. The copper-arsenic group
      a) Preliminary tests
      b) Analysis of known solution
      c) Analysis of unknown solution
   3. Group III. The aluminum-iron-nickel group
      a) Preliminary tests
      b) Analysis of known solution
      c) Analysis of unknown solution
   4. Group IV. The barium group
      a) Preliminary tests
      b) Analysis of known solution
   5. Group V. Alkali group
      a) Analysis of a known solution
      b) Analysis of unknown solution for Groups IV–V
G. Analysis of general unknown (all cations)
H. Analysis of unknown alloy
I. Analysis of single salts for cation and anion
J. Analysis of salt mixtures for cations and anions

13
Preparation of special equipment

Wash bottle. Use a 250 ml Florence flask (Figure 3) or procure a plastic wash bottle. The latter, which is convenient and inexpensive, is obtainable from most chemical supply houses.

Stirring rods and spatulas. Prepare four stirring rods by cutting two 10″ lengths of 3 mm solid glass rod. Heat each in the middle until it is soft, remove from the flame, and draw it out to about 1 mm diameter. Cut it in two pices, and reheat the small end until a small ball is formed. Heat the opposite end until a larger ball forms and, while it is hot, press it down firmly on an asbestos pad, forming a head or plunger (used for mixing solutions in small test tubes). (See Figure 4.)

Prepare four spatulas from 5″ lengths of glass rod. Heat one end of the rod until a bead forms. Remove it from the flame and pinch it flat with tongs or between other metal surfaces. Avoid getting the widened portion too wide to reach the bottom of small test tubes. Put a knob on the other end of the rod.

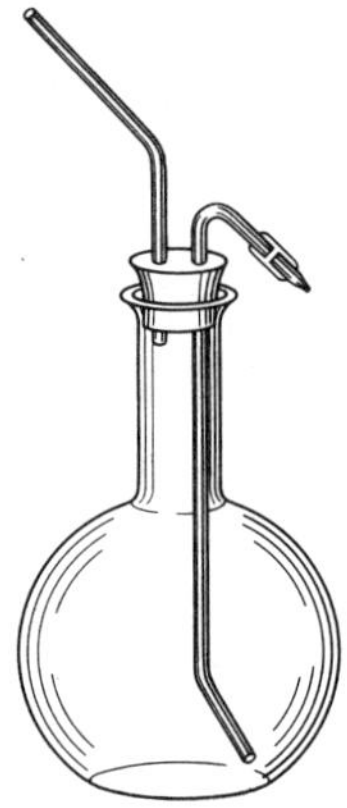

FIGURE 3

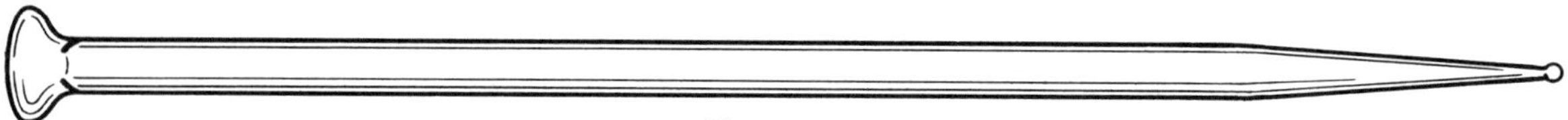

FIGURE 4

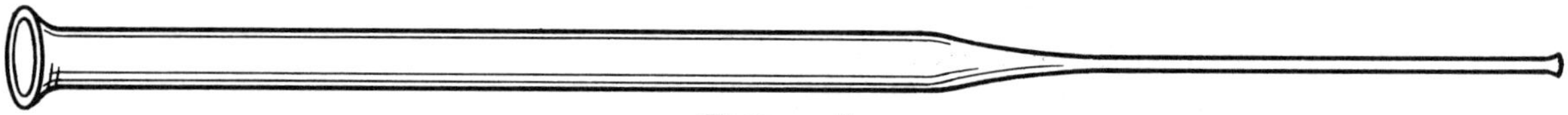

FIGURE 5

Pipets. Prepare two pipets for drawing off the liquid above a precipitate (supernatant liquid) in a small test tube after centrifuging. The pipets are like medicine droppers except that the small end should be drawn out about 2″ for easier insertion into small-diameter test tubes. (Figure 5.)

Platinum wire for flame tests. Purchase a piece of platinum wire about 2–2½″ in length. Seal the wire into one end of a piece of glass tubing. (Figure 6.)

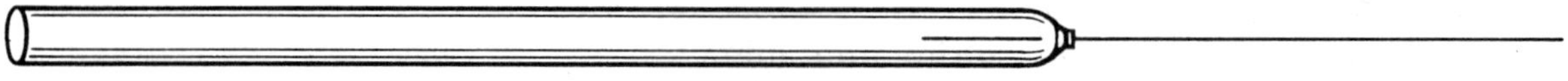

FIGURE 6

FLAME TEST. In a flame test, the substance to be tested should be a solid or a concentrated solution. If a solution is to be tested, evaporate to dryness and moisten the residue with a drop of conc. HCl. Heat the platinum wire to redness and touch it to the solid in such a way that some solid adheres to the wire. Introduce the wire again into the flame and observe the color of the flame produced. Make flame tests on solid NaCl, $BaCl_2$, $SrCl_2$, and $CaCl_2$. Before each test clean the wire by dipping it into conc. HCl in a watch glass and reheating in the flame until no color is evident. Record the colors produced.

Sodium yellow Barium green Strontium red Calcium orange

Preparation of a borax bead. Make a circular loop about 2–3 mm in diameter at the end of the platinum wire prepared above. Heat the wire to redness in the flame and quickly plunge it into powdered borax. Heat in the flame until the borax fuses to a clear colorless bead.

BORAX BEAD TEST. In making a borax bead test, heat the bead until it just softens, then touch it to a small bit (a bulk of about 1/10 that of the head of a pin) of the substance to be tested. Heat again in the oxidizing flame (above the inner blue cone of the flame). Allow to cool and observe the color. It is important that not too much substance be used in the test, since an excess may give the bead a black and opaque appearance. Make a bead test on samples of $Co(NO_3)_2$, $Ni(NO_3)_2$, and $FeCl_3$. Record the colors produced.

Cobalt__________ Nickel__________ Iron__________

14
Preliminary experiments on laboratory techniques

Laboratory technique. The purposes of this experiment are (1) to become familiar with the use and care of the centrifuge, (2) to learn the technique of handling and washing small precipitates, and (3) to emphasize the value of careful techniques.

Place 1 drop of Hg_2^{++} test solution (t.s.) and 1 drop of Ba^{++} t.s. in a test tube. Add 10 drops of water and stir the solution. Add 2 drops dil. HCl (1 *M*) and note the precipitation of mercurous chloride. Mix by shaking the tube or, better, by stirring with the "plunger" stirring rod prepared earlier. Centrifuge the mixture after carefully balancing it with another tube containing water. Fifteen to twenty seconds of centrifuging is long enough to settle most precipitates. Carefully add another drop of dil. HCl to the supernatant liquid to test for complete precipitation.

Draw off the supernatant liquid (decantate or filtrate) with one of the pipets you have prepared. Place the solution in a test tube, neutralize with dil. NH_4OH, and add a drop of K_2CrO_4 solution. The yellow precipitate is $BaCrO_4$, and shows that barium ion remains in solution after centrifuging. Wash the residue obtained above by stirring with 10 drops of water. Centrifuge and test the washings again by addition of K_2CrO_4 to see if any barium ion remains. Repeat the washings until barium ion is no longer found. How many washings were necessary to remove Ba^{++}?

Write ionic equations for the reactions taking place in the above procedure.

The use of thioacetamide as a source of H_2S. The formation of insoluble sulfides is the basis of a considerable portion of cation analysis. The source of sulfide ions for the precipitations is H_2S, which is most conveniently produced from thioacetamide. This substance hydrolyzes quite readily in water at 70°–90° C, but remains relatively stable in aqueous solution at room temperature.

$$CH_3CSNH_2 + 2\ H_2O = CH_3COONH_4 + H_2S$$

An 8% solution of thioacetamide may be bottled as a laboratory reagent. When H_2S is needed in semimicro qualitative precipitations, satisfactory results are obtained by adding 6 drops of the 8% thioacetamide to the cation solution in a test tube, which is then placed in a boiling water bath for 5 minutes (repeated once or twice if necessary). In adding the thioacetamide solution, use a glass stirring rod to ensure mixing. The ammonium acetate formed in the reaction has some merit in aiding the coagulation of the precipitate.

Place 1 ml of test solutions of Cu^{++}, As^{3+}, Sb^{3+}, Zn^{++}, Mn^{++}, Ca^{++}, and K^+, respectively, in small test tubes. Add 1 ml of water and 5 drops of 1 *M* HCl to each tube. The solution in each tube is now to be saturated with H_2S. Add 6 drops thioacetamide solution to each tube, and place the tubes in a boiling water bath as described above.

To the test tubes containing cations where no precipitation took place, add dil. NH_4OH dropwise to basicity. What 2 cations now precipitate as sulfides in basic solution? Record the colors of all the precipitated sulfides. What two cations of those tested do not form water-insoluble sulfides? (Use space below for recording results.)

A preliminary observation from this experiment may be that certain cations will precipitate as sulfides in the presence of acid, some precipitate as sulfides only in neutral or basic solution, and others do not precipitate from water solution as sulfides. This generalization is to be made use of in chemically testing an unknown cation solution. A cation unknown can at least be narrowed down as belonging to one of three groups according to its action with H_2S (thioacetamide).

Part three

Analysis for anions

15
Introduction to analysis for anions

Negatively charged ions, whether monatomic, like the halide ions, or radicals, such as the sulfate and nitrate ions, are called *anions*, from the fact that they migrate to the anode during electrolysis. Similarly, positively charged ions are called *cations* because they migrate toward the cathode or negative pole during electrolysis. The analysis of a sample divides itself into two sections virtually independent of each other; one is concerned with the identification of cations, the other with the identification of anions. In this manual, anion analysis will be considered first for two reasons: (1) many of the tests for anions have been recently made in general chemistry and hence the student is already familiar with a section of anion analysis; (2) the tests for anions are relatively simple and may be carried out with a minimum of interference from other ions present. Although a systematic procedure for separation and identification of anions may be developed, such a procedure is tedious and time-consuming. In general, anions are adaptable to identification tests which are quickly and simply carried out.

After one becomes thoroughly familiar with procedures for testing for both cations and anions, it is usually advantageous to test for cations first, since, as will be pointed out later, elimination of certain anions may be indicated on the basis of the cation analysis.

This manual will describe tests for the following anions:

- Sulfate, SO_4^{--}
- Sulfite, SO_3^{--}
- Carbonate, CO_3^{--}
- Borate, BO_3^{3-}
- Chromate, CrO_4^{--}
- Arsenate, AsO_4^{3-}
- Phosphate, PO_4^{3-}
- Chloride, Cl^-
- Bromide, Br^-
- Iodide, I^-
- Sulfide, S^{--}
- Nitrate, NO_3^-
- Nitrite, NO_2^-
- Acetate, $C_2H_3O_2^-$

It is customary to apply the following indicative tests in the course of analysis for anions: (1) action of conc. H_2SO_4 on the sample (if a solid), (2) solubility group tests on a solution of the sample, and (3) specific or individual tests.

To familiarize students with these procedures, sodium salts of the anions listed above are available in the laboratory for the tests (except that ZnS or CaS is preferable to the sodium salt). Before you proceed to the analysis of the unknown, apply the three tests above to each of the known sodium salts. (If desired, the class may be provided with a set of test solutions of anions of the sodium salts for making group tests and certain of the specific tests.) Follow Procedures 1, 2, and 3 in making the above three tests. (See pages 69, 71, and 77.)

16
Action of concentrated sulfuric acid on salts

A great deal of information can be obtained concerning the character of the acidic radical of a compound by heating moderately with a high-boiling acid (conc. H_2SO_4 or H_3PO_4), which produces the acid of the unknown anion. This acid may decompose or be oxidized, and a gas may be evolved which can be identified by color or odor. A small amount of solid material (0.1 g or less) should be used for this test. Avoid breathing any unknown gas deeply, and never point the heated tube at your neighbor.

Hot conc. H_2SO_4 is very corrosive, and CAUTION should be exercised in the heating outlined in the procedure below. Follow carefully the directions for the amounts to be used. Raise the temperature *gradually* while looking for evidence of chemical change.

Apply Procedure 1 to each of the sodium salts of the anions listed on page 68.*

PROCEDURE 1 *Action of Sulfuric Acid on Sample*

Place about 0.1 g of solid salt (about as much as can be heaped on 1/8″ of the end of the spatula) in a small test tube and moisten with a few drops of conc. H_2SO_4. Warm gently. DO NOT BOIL! Indications of possible anions present are summarized in the table below.

Result of test	*Inference as to possible anion*
No change	Sulfate, Phosphate, Borate, Arsenate
Sharp odor of vinegar or acetic acid	Acetate
Frothing and sharp odor of SO_2, effervesces Solid sulfite also emits SO_2 with cold dil. HCl.	Sulfite
Odor of rotten eggs; H_2S, S, and possibly SO_2	Sulfide (Iodide)
Brownish vapor, characteristic odor and color of bromine	Bromide (Nitrite)
Black solid, violet vapor, odor of H_2S	Iodide
Brown vapor of NO_2 (in cold), vigorous effervescence	Nitrite (Bromide)
Nitric acid, vaporizing on *high* heating One may notice a little brown NO_2 against a white background if some bit of reducing impurity is present. Otherwise, the HNO_3 colorless fumes condense on the cooler upper part of the test tube. Placing some Cu shreds in the mouth of the tube and heating, the HNO_3, as it evolves, gives characteristic action on Cu with brown NO_2 formed, and some green $Cu(NO_3)_2$.	Nitrate
Orange to red color from yellow	Chromate
Effervescence, sharp odor of HCl	Chloride
Vigorous effervescence, colorless, odorless gas	Carbonate

On page 70, write balanced equations for the reaction of sulfuric acid on these salts.

* Instructors may choose to delete testing for certain anions, such as arsenate, nitrite, and chromate.

17
Solubility group tests

The anions are grouped as follows:

Group I. The sulfate group: SO_4^{--}, CO_3^{--}, CrO_4^{--}, AsO_4^{3-}, PO_4^{3-}, SO_3^{--}, BO_3^{3-}

Group II. Halide group: Cl^-, Br^-, I^-, S^{--}

Group III. Nitrate group: NO_3^-, NO_2^-, $C_2H_3O_2^-$

This division of the anions is based on the following facts: (1) In ammoniacal solutions only the anions of Group I are precipitated with Ba^{++}; (2) in HNO_3 solution only the anions of Group II are precipitated with Ag^+. Accordingly, the group reagent for Group I is NH_4OH and $BaCl_2$, the group reagent for Group II is HNO_3 and $AgNO_3$, and there is no group reagent for Group III, since all nitrates, nitrites, and acetates are soluble.

Prepare (or obtain from side shelf if test solutions of anions have been provided) 1 ml aqueous solutions of the sodium salts of all the anions mentioned above. (Dissolve a few crystals, a volume about that of the head of a safety match, of the salt in 1 ml H_2O. The solutions used in making the group tests and specific tests must not be too concentrated, since some of the tests may be invalidated in concentrated solution.) Apply Procedure 2 to each solution and record results in the table on the following page. Record the color of the precipitates.

PROCEDURE 2 *Group Tests for Anions**

SULFATE GROUP. To 5 drops of prepared solution add dil. NH_4OH (1) until basic, then 2 drops $BaCl_2$ solution. Formation of ppt. indicates anions of this group. If no ppt. is formed immediately, allow to stand for a few minutes, since barium borate forms slowly (2), (3). Now add conc. HCl drop by drop until the solution is acid. If ppt. dissolves completely, SO_4^{--} is absent; otherwise, SO_4^{--} is present.

HALIDE GROUP. To 5 drops of prepared solution, add 5 drops dil. HNO_3 (4). Heat, and then add 1 drop $AgNO_3$ solution. Formation of ppt. indicates anions of this group. (If a sulfite is present, silver sulfite may precipitate; if so, it is decomposed to SO_2 by heating with HNO_3.)

NITRATE GROUP. If no ppt. is obtained above, anions of this group are probably present.

* Numbers in parentheses refer to Notes (below).

NOTES.

1. The solution must be neutral or basic, since the barium salts of most of the sulfate group are soluble in acid solution and would not ppt.
2. The absence of a ppt. does not definitely eliminate borate. Always make a specific test for borate.
3. Since BaS is slightly soluble in water, a cloudiness may result as Ba^{++} is added to a sulfide.
4. If the solution is not made acid with HNO_3, some Group I anions (if present) would ppt. as silver salts.

Place an X in the proper group as indicated by the precipitates obtained in solubility group tests according to Procedure 2.

Anion	*Sulfate group (Insoluble barium salt group)*	*Halide group (Insoluble silver salt group)*	*Nitrate group (Soluble group)*
Sulfate			
Borate			
Phosphate			
Chloride			
Bromide			
Sulfite			
Arsenate			
Carbonate			
Sulfide			
Iodide			
Nitrite			
Nitrate			
Acetate			
Chromate			

18
Specific tests for anions

Following the preliminary tests on a sample for anions (that is, H_2SO_4 treatment if the sample is a solid and group tests according to Procedure 2), specific or individual tests are made for those anions which have not been eliminated as possibilities. The chemistry involved and the basis for these specific tests are summarized below. See Procedure 3 for specific directions.

Sulfate. Since $BaSO_4$ is the only barium salt (of the anions considered here) insoluble in HCl, the addition of Ba^{++} ($BaCl_2$) and HCl to a solution yields a precipitate of $BaSO_4$ if SO_4^{--} is present.

$$Ba^{++} + SO_4^{--} \rightarrow \underset{\text{white}}{\underline{BaSO_4}}$$

Arsenate. If AsO_4^{3-} ion is present, arsenic should be found in the cation analysis, since H_2S reacts with this ion to give As_2S_5.

$$2\ AsO_4^{3-} + 6\ H^+ + 5\ H_2S \rightarrow As_2S_5 + 8\ H_2O$$

The specific test for AsO_4^{3-} depends on the precipitation of white crystalline $MgNH_4AsO_4$ from a solution containing AsO_4^{3-} with magnesia mixture ($MgCl_2$, NH_4Cl, NH_4OH).

$$Mg^{++} + NH_4^+ + AsO_4^{3-} \rightarrow \underset{\text{white}}{\underline{MgNH_4AsO_4}}$$

Since PO_4^{3-} gives a similar precipitate ($MgNH_4PO_4$) with magnesia mixture, AsO_4^{3-} is confirmed by treatment of the precipitate (either $MgNH_4AsO_4$ or $MgNH_4PO_4$) with acetic acid and silver nitrate solution. If arsenate ion is present, a chocolate-colored precipitate of silver arsenate is obtained.

$$AsO_4^{3-} + 3\ Ag^+ \rightarrow \underset{\text{chocolate}}{\underline{Ag_3AsO_4}}$$

If phosphate ion is present, a yellow precipitate of Ag_3PO_4 is obtained.

$$PO_4^{3-} + 3\ Ag^+ \rightarrow \underset{\text{yellow}}{\underline{Ag_3PO_4}}$$

Phosphate. The addition of ammonium molybdate and HNO_3 to a solution containing phosphate ion yields a finely divided canary-yellow precipitate of ammonium phosphomolybdate.

$$PO_4^{3-} + 12\ (NH_4)_2MoO_4 + 21\ HNO_3 + 3\ H^+ \rightarrow \underset{\text{yellow}}{\underline{(NH_4)_3PO_4 \cdot 12MoO_3}} + 21\ NH_4NO_3 + 12\ H_2O$$

Since arsenate ion gives a similarly colored yellow precipitate of ammonium arseno-molybdate, any arsenic must be removed before the test. This may be accomplished by its precipitation as As_2S_5 with H_2S.

Borate. If conc. H_2SO_4 and methyl alcohol (or ethyl alcohol) are added to a borate, methyl borate is produced. When burned this gives an immediate greenish colored flame.

$$Na_2B_4O_7 + H_2SO_4 + 5\ H_2O \rightarrow Na_2SO_4 + 4\ H_3BO_3$$

$$H_3BO_3 + 3\ CH_3OH \rightarrow (CH_3)_3BO_3$$

$$2\ (CH_3)_3BO_3 + 9\ O_2 \rightarrow 6\ CO_2 + B_2O_3 + 9\ H_2O$$

Caution: Avoid breathing fumes of methyl borate or methyl alcohol.

Chromate. If this ion is present, chromium will be found in the cation analysis. Chromates are yellow, and this often aids in their identification. The specific test for the ion depends on the precipitation of $BaCrO_4$ in acetic acid solution.

$$Ba^{++} + CrO_4^{--} \rightarrow \underset{\text{yellow}}{\underline{BaCrO_4}}$$

Chromates in acid solution turn to an orange-colored dichromate.

$$2\ CrO_4^{--} + 2\ H^+ \rightarrow H_2O + Cr_2O_7^{--}$$

If a few drops of H_2O_2 and ether are shaken into a dilute slightly acid chromate solution, a blue color develops in the ether layer. This is probably because of the formation of a perchromic acid.

Sulfite. The test for sulfite ion is based on its oxidation to SO_4^{--} with H_2O_2, followed by precipitation of $BaSO_4$ in HCl solution.

$$SO_3^{--} + H_2O_2 \rightarrow SO_4^{--} + H_2O$$

$$SO_4^{--} + Ba^{++} \rightarrow \underline{BaSO_4}$$

If sulfate ion is present, it must be removed by precipitation as $BaSO_4$ before H_2O_2 is added. Furthermore, a solid sulfite will emit SO_2 when treated with dil. HCl.

Carbonate. Carbonates effervesce rapidly on the addition of HCl to give colorless, odorless CO_2. Effervescence in itself is a good indication of the presence of a carbonate, but, since sulfides and sulfites also effervesce with HCl, this is not a specific test. To confirm the carbonate ion, the gas evolved on addition of HCl is allowed to react with $Ba(OH)_2$ or $Ca(OH)_2$ solution, which yields a white precipitate of $BaCO_3$ or $CaCO_3$.

$$CO_3^{--} + 2\ H^+ \rightarrow CO_2 + H_2O$$

$$CO_2 + Ba(OH)_2 \rightarrow \underline{BaCO_3} + H_2O$$

Since SO_2 may give a white precipitate of $BaSO_3$ with barium hydroxide, sulfite ion must be removed if present before testing for carbonate. This may be accomplished by oxidation of sulfite ion to sulfate ion with sodium peroxide (or hydrogen peroxide).

$$2\ SO_3^{--} + 2\ Na_2O_2 + 2\ H_2O \rightarrow 2\ SO_4^{--} + 4\ Na^+ + 4\ OH^-$$

Sulfide. Sulfide ion is identified by liberating H_2S, which will react with a solution of lead acetate or lead nitrate to produce a black precipitate of PbS.

$$Pb^{++} + H_2S \rightarrow \underline{PbS} + 2\ H^+$$

Chloride, bromide, and iodide. These ions form insoluble silver halide salts with a solution of silver nitrate and nitric acid. Although the colors of the silver salts (AgCl, white; AgBr, pale yellow; and AgI, yellow) are indicative, confirmatory tests are desirable. Bromide and iodide ions may be confirmed by displacement reactions with free chlorine. The color of the liberated halogen is intensified with the use of a solvent such as carbon tetrachloride or carbon disulfide.

$$2\ Br^- + Cl_2 \rightarrow Br_2 + 2\ Cl^-$$

$$2\ I^- + Cl_2 \rightarrow I_2 + 2\ Cl^-$$

Nitrate and nitrite. If a nitrate is treated with conc. H_2SO_4 followed by careful addition of a solution of ferrous sulfate, a brown ring, nitrosyl ferrous sulfate ($FeSO_4 \cdot NO$), is formed at the junction of the sulfuric acid and ferrous sulfate solutions. Sulfuric acid liberates nitric acid from the nitrate present. The HNO_3 is reduced by ferrous sulfate—then the unstable complex $FeSO_4 \cdot NO$ is formed.

$$2\ HNO_3 + 6\ FeSO_4 + 3\ H_2SO_4 \rightarrow 3\ Fe_2(SO_4)_3 + 2\ NO + 4\ H_2O$$

$$FeSO_4 + NO \rightarrow \underset{\text{nitrosyl ferrous sulfate}}{FeSO_4 \cdot NO}$$

The test for a nitrite is similar to the test for nitrate except that acetic acid is used instead of sulfuric acid. The test for a nitrate would also yield a brown ring test with a nitrite; consequently, if a sample may contain both nitrite and nitrate, further tests would be necessary to confirm a nitrate in the presence of a nitrite.

Acetate. The addition of conc. H_2SO_4 and ethyl alcohol to an acetate, followed by warming, produces ethyl acetate, a volatile ester which is identified by its sweet and fruity odor. Sulfuric acid reacts with the acetate to give acetic acid, which then reacts with the alcohol as follows:

$$HC_2H_3O_2 + C_2H_5OH \rightarrow \underset{\text{ethyl acetate}}{C_2H_3O_2C_2H_5} + H_2O$$

An alternate test consists of testing an unknown solution with lanthanum nitrate and potassium iodide–iodine solutions. If an acetate is present, a basic lanthanum acetate is formed which will adsorb iodine to give a dark blue color. Furthermore, a dilute, slightly yellow solution of $FeCl_3$ will turn reddish when gently heated with an acetate.

Preliminary experiment on specific tests. Carry out specific tests for all the anions listed above according to Procedure 3. Use test solutions of the anions or the solid salts, whichever are called for in the procedure. If there is any question about a test working out satisfactorily, consult the instructor. If solutions are not already made up, dissolve a small amount (volume of a small pea) of the sodium salt in 2 ml H_2O. Make notes on each specific test and write equations on the left-hand blank pages.

PROCEDURE 3 *Specific Tests for Anions**

1. **SULFATE.** To 5 drops of solution, add 2 drops 6 *M* HCl and 2 drops $BaCl_2$ solution. A white ppt. ($BaSO_4$) proves SO_4^{--}.

2. **ARSENATE.** Arsenic is also determinable in cation analysis. To 5 drops of the prepared solution add 6 *M* NH_4OH until basic; then add 3 or 4 drops of magnesia mixture. If a white crystalline ppt. forms, it may be either $MgNH_4PO_4$ or $MgNH_4AsO_4$. Cfg. the ppt. Wash ppt. twice with 4 drops H_2O; then add 1 drop acetic acid and 1 drop silver nitrate solution to the ppt. Formation of a yellow ppt. (Ag_3PO_4) indicates PO_4^{3-}; formation of a chocolate-colored ppt. proves AsO_4^{3-}.

3. **PHOSPHATE.** Arsenates interfere with the following phosphate test. Unknown mixtures of phosphate and arsenic compounds will not be given. After proving arsenate absent (2 above), proceed. To 5 drops of prepared solution add 5 drops dil. HNO_3 and 5 drops ammonium molybdate solution. Heat to boiling for several seconds, and allow to stand for a few minutes. Formation of a finely divided canary-yellow ppt. [$(NH_4)_3PO_4 \cdot 12MoO_3$] proves PO_4^{3-}.

4. **BORATE.** To be made on original sample. To about 0.1 g of the solid add a few drops conc. H_2SO_4 in an evaporating dish. Mix thoroughly; add 2 or 3 ml methyl alcohol (or ethyl alcohol), heat, and then set fire to alcohol. The immediate appearance of a green flame proves BO_3^{3-}.

5. **CHROMATE.** Chromium will also be found in cation analysis. To 5 drops of prepared solution add dil. NH_4OH until basic, then dil. acetic acid until acid. Add 2 drops $BaCl_2$ solution. Formation of a yellow ppt. ($BaCrO_4$) proves CrO_4^{--}. To confirm CrO_4^{--}, cfg., and wash the ppt. twice with 6 drops cold H_2O. To the ppt. add 2 drops dil. HNO_3 and 10 drops of H_2O; then add 4 or 5 drops ether and 2 drops of 3% H_2O_2 solution. Shake tube well. A blue color appearing in the ether layer confirms CrO_4^{--}.

* If these tests are being made on sodium salts, it is only necessary to dissolve the salt in water for preparing a solution. If cations other than Na^+, K^+, and NH_4^+ are present, it will be necessary to treat the sample according to Procedure 15 for preparing a solution for anion testing.

6. SULFITE. To a small amount of solid in a small test tube, add 5 drops 6 *M* HCl, and heat gently. Effervescence and an odor of SO_2 indicates sulfite. (Do not mistake HCl odor for that of SO_2.) To further test, add dil. HCl to 5 drops of solution or to a small amount of solid until distinctly acid, dilute with several drops H_2O, then add 2 drops $BaCl_2$. If a ppt. forms, cfg. and discard ppt. ($BaSO_4$). Add 4 drops 3% H_2O_2 to the filtrate. Formation of a white ppt. ($BaSO_4$) proves SO_3^{--}.

see notebook☆

7. CARBONATE. To be made on original sample. (A) If sulfite is absent: To a small amount of solid (volume of a small pea) in test tube add a few drops 6 *M* HCl. Test for evolution of CO_2 by using a small test tube generator. Fit a small test tube with a cork carrying a delivery tube which can be dipped into a $Ba(OH)_2$ solution in another small test tube. Add a few drops conc. HCl to about $\frac{1}{4}''$ of solid in the generator, quickly replace the stopper, and allow the evolved gas to bubble into $Ba(OH)_2$ solution. A white precipitate soon forms if carbonate is present. Any sample which effervesces with HCl to produce a colorless, odorless gas must contain a carbonate.
(B) If sulfite is present: To a small amount of solid in a test tube add an equal volume of Na_2O_2; then add a few drops of H_2O. Shake thoroughly. Add dil. HCl unit acid, and proceed as above in testing for carbonate.

8. SULFIDE. To be made on original sample. To a small amount of solid in a test tube add 5 drops dil. 6 *M* HCl. Hold a strip of filter paper which has been moistened with lead nitrate- (Pb acetate) solution over the mouth of the tube. If the paper is blackened (PbS), S^{--} is present. If the sample fails to dissolve by this treatment, add a few granules of zinc to the tube and warm. Test as above with lead nitrate paper.

9. CHLORIDE, BROMIDE, AND IODIDE. To 5 drops of prepared solution add 10 drops dil. HNO_3. Heat until there is no evidence of effervescence. (If a sulfite is present, it must be destroyed before you proceed.) Add 1 drop silver nitrate solution. A white ppt. soluble in NH_4OH indicates Cl^-; a pale yellow ppt. (AgBr) sparingly soluble in NH_4OH indicates Br^-; a yellow ppt. (AgI) insoluble in NH_4OH indicates I^-. To detect bromide or iodide, add 6 drops chlorine water to 5 drops of prepared solution, followed by 10 drops CCl_4. Shake the tube well. The CCl_4 layer will be colored brown by bromine or violet by iodine. (To test for the presence of Cl^-, Br^-, and I^- in the presence of one another, see notes.)

10. NITRATE. To be made on original sample. If unknown is a solid the test for a nitrate as on page 69 is conclusive, and further tests need not be made. (A) In absence of bromide or iodide: Dissolve a pinch of the solid in a few drops of H_2O, or, in the event the sample is insoluble in H_2O, add 3 drops 6 *M* HCl. If there is any residue after acid treatment, cfg. and discard. Add 10 drops conc. H_2SO_4 to solution in test tube, then cool tube under tap. Now incline the tube at an angle of about 30° with the horizontal and add several drops of freshly prepared $FeSO_4$ solution *very slowly*, so that it forms a layer which floats on the H_2SO_4 solution. (The $FeSO_4$ solution may be prepared by dissolving 1 g of solid in 1 or 2 ml H_2O.) Bring the tube to an upright position and allow it to stand for about 1 minute. The formation of a brown ring ($FeSO_4 \cdot NO$) at the junction of the solutions proves NO_3^-.
(B) In presence of bromides or iodides: Prepare a solution for the nitrate test as above; then add silver sulfate solution until pptn. is complete. Cfg. and discard ppt. (AgBr, AgI). Test the filtrate as above with $FeSO_4$ and H_2SO_4.

11. NITRITE. To be made on original sample. Treat a small amount of the solid with a few drops H_2O. Reject any residue. Test the solution as in (10), but use acetic acid instead of H_2SO_4. A deep brown color proves NO_2^-.

12. ACETATE. To be made on original sample. Treat a small amount of solid with 6 drops conc. H_2SO_4, followed by addition of 6 drops ethyl alcohol. Warm tube slowly to near boiling. A sweet or fruity odor (ethyl acetate) proves $C_2H_3O_2^-$.

Alternate test 1. (A) In absence of sulfate group: Dilute 2 or 3 drops of unknown solution to

about 1 ml or, if sample is a solid, dissolve a small amount in H_2O. Add 4 drops $La(NO_3)_3$ solution (5%). If ppt. forms, add 1 or 2 drops dil. HNO_3 to redissolve. Add 1 drop $KI-I_2$ solution, make just alkaline with 6 *M* NH_4OH, and warm in water bath. A deep blue-black ppt. indicates acetate ion.
(B) If members of sulfate group are present, they must be removed. Dilute 2 or 3 drops of unknown solution to about 1 ml. Add 1 or 2 drops $BaCl_2$ solution and 6 *M* NH_4OH until alkaline. Filter off any ppt. and reject it. Treat solution according to (A).

Alternate test 2. Use 1 ml of dil. yellow $FeCl_3$ solution and add a small crystal of unknown solid or 4 drops of solution to it. Heat gently. A reddish color to the solution indicates acetate (3).

NOTES.

1. Some instructors may want to add tartrate as an example of an organic compound. In general, organic compounds char when heated with conc. H_2SO_4.

TARTRATE. (a) To a small quantity of $Na_2C_4H_4O_6$ in a small test tube, add a few drops of conc. H_2SO_4. Heat gently, and note charring and some evolution of gas (SO_2).
(b) Silver mirror test. Clean a 6″ test tube by boiling NaOH solution in it and then washing it thoroughly with H_2O. Add 2 ml $AgNO_3$ to the tube; then add dil. NH_4OH *drop by drop* with shaking until the ppt. which forms just dissolves or almost dissolves. Avoid excess NH_4OH (it is best to stop adding NH_4OH when a slightly gray ppt. of AgOH is still apparent). Add about 3 ml of the unknown solution and warm gently. Tartrates, if present, will reduce the silver compound to metallic silver, which will coat the inside of the test tube as a mirror.

$$AgNO_3 + NH_4OH \rightarrow AgOH\downarrow + NH_4NO_3$$

$$AgOH \rightarrow Ag^+ + OH^-$$

$$Ag^+ + 2\ NH_3 \rightarrow Ag(NH_3)_2^+$$

$$Ag(NH_3)_2^+ + Na_2C_4H_4O_6 \rightarrow Ag + 2\ NH_3 + \text{oxidation product}$$

2. Detection of Cl^-, Br^-, I^- in the presence of one another: In the brief scope of this course unknown mixtures of halogen salts will not be given. The literature has many references on methods of separation of the halogens, and thus separate identification. Tests given in the past few pages are adequate for the detection of iodide or bromide in chloride. An example of a method of removing bromine and iodine from chloride for follow-up chloride detection is as follows: A water solution of the three halide salts may be boiled with a few crystals of potassium dichromate. Iodine is liberated; on adding a few drops of H_2SO_4, Br_2 is liberated.

$$6\ I^- + Cr_2O_7^{--} + 7\ H_2O \rightarrow 2\ Cr(OH)_3 + 3\ I_2 + 8\ OH^-$$

$$Cr_2O_7^{--} + 6\ Br^- + 14\ H^+ \rightarrow 2\ Cr^{3+} + 3\ Br_2 + 7\ H_2O$$

Heating or CCl_4 extraction may be used to remove the free Br_2 or I_2 from the solution, which would still contain chloride ion.

3. The *solution* should be red here. If anions are present which form insoluble ferric salts, a reddish rust-colored precipitate will be obtained, for example, CO_3^{--} would precipitate ferric carbonate, etc.

Analysis of anion unknowns. After completion of (1) action of conc. H_2SO_4, (2) group tests, (3) specific tests on the sodium salts of the various acid radicals, and (4) approval of written work on these, the instructor will issue to students a group of five unknowns which are to be tested for anions.

(For anion unknowns, it is suggested that groups of five corked small bottles or test tubes be bound together with a rubber band. A different solid sodium salt is to be placed in each bottle or tube. Numbers of tubes may be printed on the corks. The instructor should have sets of such unknowns prepared ahead of time. Since sodium sulfide is hygroscopic, use CaS or ZnS.)

First try the action of conc. H_2SO_4, Procedure 1. If this gives a good idea as to the nature of the anion, proceed immediately to specific tests, Procedure 3. If no good clues are obtained in the H_2SO_4 treatment, analyze according to Procedures 2 and 3. On the opposite page, record observations on all tests made on the unknowns. Each record should lead to conclusions as to the anion present. Present this written report to your instructor.

Analysis of anion mixtures. When analyzing an unknown which contains two or more anions, in general, the procedure given above should be followed. The action of sulfuric acid on the mixture should indicate a great deal. Follow with solubility group tests and specific tests.*

Be sure to confirm evidence which suggests certain anions. For example, treatment of an iodide with concentrated H_2SO_4 suggests sulfite and sulfide as well as iodide, since the later reduces H_2SO_4 to SO_2 and H_2S. However, if S^{--} and SO_3^{--} are absent in the original sample, this will be indicated by confirmatory tests for these anions.

* It is true that certain anions interfere with the testing for others, but it is beyond the scope of this course to consider all such problems. Instructors should issue anion mixtures which will be compatible with the outlined procedures of analysis.

Part four

Analysis for cations

Group Separations*

Ag^{+}, Pb^{++}, Hg_2^{++} — Group I

Hg^{++}, Cu^{++}, Bi^{3+}, Cd^{++}, As^{3+}, AsO_4^{3-}, Sb^{3+}, SbO_4^{3-}, Sn^{++}, Sn^{4+} — Group II

Fe^{++}, Fe^{3+}, Al^{3+}, Cr^{3+}, Zn^{++}, Co^{++}, Ni^{++}, Mn^{++} — Group III

Ba^{++}, Sr^{++}, Ca^{++} — Group IV

Mg^{++}, Na^{+}, K^{+}, Li^{+}, NH_4^{+} — Group V

[HCl]

- Group I chlorides
- Groups II, III, IV, V
  - [HCl, H_2S]
  - Group II sulfides
  - Groups III, IV, V
    - [NH_4OH, NH_4Cl, H_2S].
    - Group III sulfides and hydroxides
    - Groups IV, V
      - [$(NH_4)_2CO_3$, NH_4OH, NH_4Cl]
      - Group IV carbonates
      - Group V

* In the above diagrammatic representation of the cation group separations,

——— a horizontal line indicates a filtration or separation by centrifuge

╓—— denotes a precipitate

——┐ denotes a filtrate

[ ] denotes an added reagent

19

Introduction to cation analysis

Among the problems confronting the analyst are (1) identification of metals, (2) analysis of alloys for metallic components, and (3) detection of metals as constituents of compounds. In the case of water solutions of compounds of metals, the problem is one of *identification of metallic ions* (cations), which is the text of this section of the manual. Dissolving metals and alloys to obtain ions and the solution of compounds insoluble in water will be discussed later. In the brief course of qualitative analysis for which this manual is designed, only the more common or representative cations are included for analysis.

Method of cation analysis. Some few cations may be detected in the presence of several other cations, but the usual procedure for analyzing a mixture of several cations involves a *systematic separation of cations into groups, followed by separation of each group into its components.* By taking advantage of differences in chemical properties, groups of ions are separated from other ions in solution through *group reagents* which form precipitates with given groups. To illustrate, if hydrochloric acid is added to a solution containing all cations, only the chlorides of lead, silver, and mercury (*ous*) precipitate, since all other chlorides are soluble. (See solubility rules, page 59.) Lead, silver, and mercury (*ous*) ions then constitute a group of ions which may be precipitated from solution by addition of the group reagent HCl. After removal of this group of ions, another group may be precipitated by adding hydrogen sulfide to the acidified solution. By treating the resulting solution, other groups may be separated by addition of group reagents; that is, the cations are divided into five groups based on differences in chemical properties:

Group I. Ag^{+}, Pb^{++}, Hg_2^{++}

Group II. Pb^{++}, Hg^{++}, Cu^{++}, Bi^{3+}, Cd^{++}, As^{3+}, AsO_4^{3-}, Sb^{3+}, SbO_4^{3-}, Sn^{++}, Sn^{4+}
(Arsenic and antimony probably do not exist as simple ions when exhibiting oxidation numbers of 5.)

Group III. Fe^{++}, Fe^{3+}, Al^{3+}, Cr^{3+}, Ni^{++}, Co^{++}, Mn^{++}, Zn^{++}

Group IV. Ba^{++}, Sr^{++}, Ca^{++}

Group V. Mg^{++}, Na^{+}, K^{+}, Li^{+}, NH_4^{+}

The separation of the five groups is shown diagrammatically on page 86.

After the ions have been separated into groups, it is necessary to separate the ions of each group; these separations are also dependent on differences in chemical properties of the constituent ions. For example, in Group I, lead chloride may be separated from silver and mercurous chlorides by taking advantage of the fact that lead chloride is soluble in hot water, whereas the chlorides of mercury and silver are relatively insoluble. The chlorides of mercury and silver then can be separated with ammonium hydroxide, since silver chloride is soluble and mercurous chloride is insoluble in that reagent. After the ions have been separated from one another, *specific tests* are applied; that is, a reagent is added which gives a characteristic precipitate or color with the ion in question.

In general, the procedure for each of the cation unknowns will be: initially, a group separation is made, then follows a separation of the ions in each group, and finally, after separation, specific tests are made for the cations of each group. The chemical basis for each of the group separations and specific tests will be considered at the appropriate place in the qualitative procedure.

The diagrammatic scheme of analysis shown on page 86 should be studied carefully, for similar diagrammatic schemes are to be developed as each sample is analyzed by the student. **The principal purpose of such an outline is a record of steps taken in the analysis. Each step taken in the analytical procedure is to be recorded in this outline at the time of doing the experimental work. Colors of precipitates and colors of solutions should be carefully noted on this outline, since these colors are often indicative of the presence of certain ions in the sample.** Before any sample is reported, a careful check of this outline should be made to see that the ions to be reported conform in properties to the notes made. **Such a schematic representation of the analysis will be required with every report of analysis.**

In general, the student will first analyze a known solution (that is, a solution known to contain certain specific ions) before proceeding to the analysis of an unknown sample. The purpose of the method is to familiarize the student thoroughly with the techniques involved in the group separations and final tests before applying them to an unknown. A series of preliminary tests for the individual ions precede each of the group analyses. These may be performed at the option of the instructor.

20 Preliminary experiment on group separations

This experiment illustrates the use of group reagents in separating the cations into five major divisions. One cation from each group has been chosen, and a simple procedure is given for each separation. As the procedure is carried out, the student should develop a diagrammatic scheme for the steps similar to the one on page 86. Use page 88 for the scheme, and write ionic equations for the precipitation of the cations.

Preliminary Experiment

Place 2 drops of each of the following test solutions in a small test tube: Ag^{+}, Cu^{++}, Ni^{++}, Ba^{++}, and K^{+}. Add 10 drops of H_2O and 2 drops of 1 *M* HCl. Stir, heat, and then cfg. A ppt. indicates an element of Group I. Test for complete pptn. Decant or pipet the solution to another tube (1).

Adjust the hydrogen ion concentration for Group II pptn. as follows: Add dil. NH_4OH drop by drop until the solution shows violet on methyl violet paper (2). (Use a wire or glass rod to wet a small spot on the paper.) Acidify the solution with 6 *M* HCl until just green to methyl violet paper. Add 2 drops of NH_4Cl solution. Saturate the solution with H_2S by adding 6 drops of thioacetamide solution. Then heat the tube in a boiling water bath for at least 5 minutes. The formation of a ppt. indicates a cation of Group II. Cfg. If the solution is not clear, bring it to a boil and cfg. again. Test for complete pptn.

Transfer the solution to a crucible and heat to boiling to remove H_2S(3). Add conc. NH_4OH until the solution is basic to litmus. Transfer the mixture to a test tube and saturate with H_2S by again adding 6 drops thioacetamide and heating in a boiling water bath for 5 minutes. Pptn. indicates a cation of Group III. Cfg.

Remove the supernatant liquid; then evaporate the solution to a volume of about 1 ml. Cfg. again if not clear. Add 2 drops NH_4Cl solution, make basic with NH_4OH, and add 2 drops $(NH_4)_2CO_3$ solution. Heat to boiling. Pptn. indicates a cation of Group IV. Cfg. The presence of a cation of Group V can be proved only by making specific tests on the filtrate.

NOTES.

1. This is a simplified procedure adapted to the 5 ions present. In a general unknown, the procedure under each group should be followed.

2. The proper acidity may also be obtained without the use of methyl violet paper in the following manner: Carefully neutralize the solution by adding 6 *M* NH_4OH dropwise until the solution tests just basic with litmus paper. Then add *exactly* 2 drops of 6 *M* HCl.

3. If the test tube is not over half full, the boiling may be done in the tube if it is heated very cautiously. If boiled in a crucible, transfer back to a test tube before proceeding.

Group I   The Silver Group

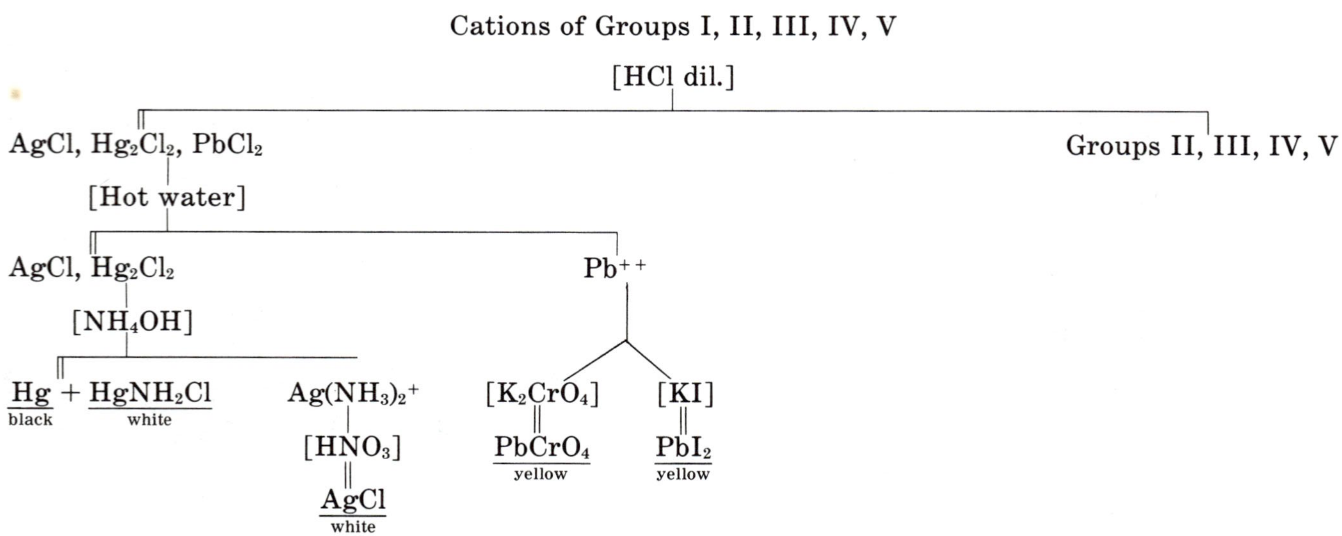

21 Group I the silver group

A. Chemistry of separations and tests.

1. The fact that chlorides of Ag^+, Pb^{++}, and Hg_2^{++} are insoluble is the basis for the separation of Group I from Groups II, III, IV, and V. Since only Group I chlorides are insoluble, addition of chloride ion (HCl) precipitates AgCl, $PbCl_2$, and Hg_2Cl_2 from a solution containing all of the cations. Refer to the diagrammatic procedure on page 90.

$$Ag^+ + Cl^- \rightarrow \underline{AgCl}$$

$$Pb^{++} + 2\ Cl^- \rightarrow \underline{PbCl_2}$$

$$Hg_2^{++} + 2\ Cl^- \rightarrow \underline{Hg_2Cl_2}$$

2. $PbCl_2$ is soluble in hot water; hence it may be separated from AgCl and Hg_2Cl_2, which remain insoluble. Lead is identified by precipitation of yellow lead chromate

$$Pb^{++} + CrO_4^{--} \rightarrow \underline{PbCrO_4}$$

or precipitation of orange-yellow lead iodide.

3. AgCl can then be separated from Hg_2Cl_2 because of its solubility in NH_4OH to form the complex soluble ion $Ag(NH_3)_2^+$, called diammine silver ion.

$$\underline{AgCl} + 2\ NH_3 \rightarrow Ag(NH_3)_2^+ + Cl^-$$

4. Hg_2Cl_2 is converted to insoluble Hg and $HgNH_2Cl$ (mercuric ammonobasic chloride) by its action with NH_3.

$$Hg_2Cl_2 + 2\ NH_3 \rightarrow \underset{\text{black}}{\underline{Hg}} + \underset{\text{white}}{\underline{HgNH_2Cl}} + NH_4^+ + Cl^-$$

In this equation one mercurous atom is oxidized to oxidation number 2, while the other is reduced to atomic mercury of zero oxidation state. The black precipitate produced serves as a test for the presence of mercurous ion.

5. Silver is identified by reprecipitation of AgCl from the ammoniacal solution with $H^+(HNO_3)$.

$$Ag(NH_3)_2^+ + Cl^- + 2\ H^+ \rightarrow \underline{AgCl} + 2\ NH_4^+$$

B. Preliminary tests for Group I. Record observations and write equations for all chemical changes on the opposite page.

1. Solubility of Group I chlorides in water.
   Place 3 drops of Pb^{++} test solution (t.s.) in a test tube and add 1 drop of 6 *M* HCl. Cool under the tap and let stand for a few minutes. Heat to near boiling and observe any changes.
   Repeat the procedure, using 1 drop of Hg_2^{++} t.s. and again with 1 drop of Ag^+ t.s. Cfg. and save ppts. for subsequent tests.

2. Precipitation of $PbCrO_4$.
   To 1 drop of Pb^{++} t.s. add 1 drop of K_2CrO_4. What change takes place?

3. Action of ammonia on Hg_2Cl_2 and AgCl.
   To the Hg_2Cl_2 and AgCl from (1) add 15 *M* NH_4OH dropwise and observe changes. To the tube which contained AgCl add 6 *M* HNO_3 until acid.*

C. Analytical procedure for Group I. To become thoroughly familiar with the separations and tests for ions of this group, the student should make up a sample containing all of the ions of the group (called a known solution) and analyze according to Procedure 4 below. For the known sample use 2 drops each of Ag^+ and Hg_2^{++} and 10 drops of Pb^{++} test solutions. Save the specific or final tests to show to the instructor.

* Throughout qualitative analysis, it is important to check acidity or basicity often. In this case, a few drops of HNO_3 might seem to be enough, but since 15 *M* NH_4OH was used, an unexpectedly larger amount of HNO_3 is needed to render the solution slightly acid. Errors are made at times by not stirring a column of liquid in a small test tube: the upper part may be acid and the lower part basic, and vice versa.

PROCEDURE 4 *Analysis of Silver Group*

SOLUTION. (1) Place 0.5 ml of the solution to be analyzed in a 3″ or 4″ test tube. Add 1 drop 6 *M* HCl (2). Heat the tube by inserting into a bath of hot water, in order to coagulate the ppt. Cool under the tap and allow to stand 3 minutes (3). Cfg. Test for complete pptn. by adding a drop of 6 *M* HCl to the supernatant liquid in the tube. Pipet or decant the solution into a test tube. Wash the ppt. with several drops 1 *M* HCl (4) and add washings to the solution already drawn off.

PRECIPITATE: AgCl, Hg_2Cl_2, $PbCl_2$. Add 8 drops H_2O, heat the tube to near boiling in a water bath, and shake to digest the ppt. (5). Cfg. Before cooling, decant the solution. Wash the ppt. with 4 drops more of hot water. Combine the washings (6).	**SOLUTION:** Groups II, III, IV, V. Analyze according to Procedure 5.
PRECIPITATE: Hg_2Cl_2, AgCl. Add 4 drops of conc. NH_4OH. Formation of a black ppt. proves mercury. Shake the tube. Cfg. Decant the solution. Repeat the extraction with 4 drops more of NH_4OH and combine the two extractions.	**SOLUTION:** Pb^{++}. To 2 drops of the solution in a test tube add 1 drop of K_2CrO_4. Formation of yellow ppt. indicates Pb^{++} (7).
PRECIPITATE: $HgNH_2Cl$, Hg.	**SOLUTION:** $Ag(NH_3)_2^+$. To the solution in a test tube add dil. HNO_3 drop by drop until the solution is definitely acid to litmus. White ppt. confirms *silver.*

NOTES.

1. A known for any group should contain about 1.0 mg of each ion. It is prepared by mixing 2 drops of test solution of each of the ions to be tested, except lead. Ten drops of Pb^{++} test solution should be used because of the relatively higher solubility of $PbCl_2$.
When an unknown solution is received from the instructor, use only 0.5 ml for analysis. Reserve the rest for repeating doubtful tests.

2. A large excess of HCl will partially dissolve AgCl and Hg_2Cl_2.

3. $PbCl_2$ is slow to ppt. If only Group I is to be tested, run a test on the filtrate for Pb^{++}.

4. 1 *M* HCl will dissolve SbOCl and BiOCl, which may have precipitated if present.

5. The use of a small beaker of water for a water bath is the safest manner of heating to boiling; however, if the volume is less than 1 ml, heating can be done directly over a micro flame.

6. If a large amount of $PbCl_2$ is present here, as might be obtained from many alloys, several washings may be necessary to completely remove $PbCl_2$. If some $PbCl_2$ remains here, it will not interfere in the test for Hg_2^{++} since, if the latter is present, a black ppt. will be obtained on addition of NH_4OH. White $PbCl_2$ remains unchanged.

7. An alternate test for lead is the formation of an orange-yellow ppt. with dil. KI solution.

Equations for Group I

Silver

Mercury

Lead

D. Analysis of Group I unknown. After completion of the known sample, write equations for all reactions involved in the procedure on page 96. Group together all reactions of each cation of Group I; these equations should be in ionic form. Submit the completed equations to your instructor, who will then quiz you on the chemistry involved in Group I procedures. After you have demonstrated familiarity with the group procedures, the instructor will issue an unknown, which may contain any or all 3 of the ions. This unknown is to be analyzed in the same manner as was the known solution, according to Procedure 4. The only write-up required for an unknown is the diagrammatic scheme of analysis which is *to be developed while making the analysis*. In this scheme, show all operations performed and all colors of solutions and precipitates. Use the space below for the diagrammatic scheme of your unknown. Report the results to your instructor.

Group II The Copper-Arsenic Group

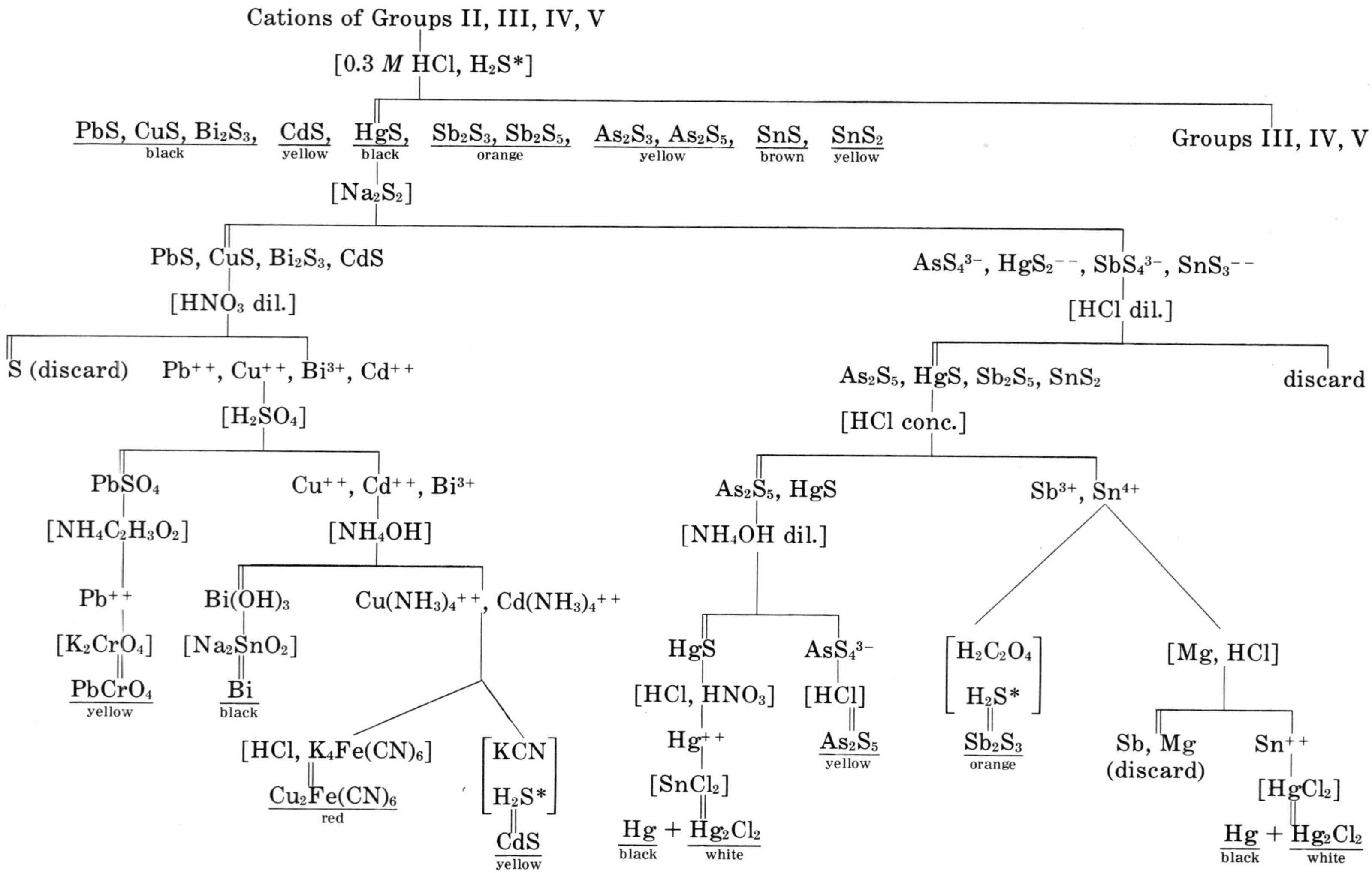

* Or thioacetamide

22
Group II
the copper-arsenic group

A. Chemistry of separations and tests.

Precipitation and Separation of Copper and Arsenic Subgroups

1. When H_2S is added to a 0.3 *M* HCl solution containing cations of Groups II, III, IV, and V, only the sulfides of Group II are precipitated, because the sulfides of cations of Groups III, IV, and V are soluble in 0.3 *M* HCl. The sulfides precipitate in approximately the following order: As_2S_5, As_2S_3, HgS, CuS, Sb_2S_3, Sb_2S_5, SnS_2, Bi_2S_3, PbS, CdS, SnS. An example of one of the precipitation reactions would be:

$$Cu^{++} + H_2S \longrightarrow \underline{CuS} + 2\ H^+$$

(Refer to page 35 for application of the solubility product principle to the precipitation of cations of Group II.)

2. The basis for the separation of Hg, As, Sb, Sn from Cu, Pb, Bi, Cd is the fact that the sulfides of the arsenic subgroup are soluble in sodium polysulfide solution (Na_2S_2), while the sulfides of the copper group are insoluble. (Refer to diagrammatic scheme on page 98.) Treatment with Na_2S_2 oxidizes any lower valence As, Sb, Sn to the high oxidation states, for example,

$$As_2S_3 + 2\ S_2^{--} \longrightarrow As_2S_5 + 2\ S^{--}$$

$$SnS + S_2^{--} \longrightarrow SnS_3^{--}$$

The polysulfide solution contains sulfide ions (S^{--}) as well as disulfide ions (S_2^{--}). The former react with the high-valence sulfides to form thio salts, for example,

$$As_2S_5 + 3\ S^{--} \longrightarrow 2\ AsS_4^{3-}$$

$$SnS_2 + S^{--} \longrightarrow SnS_3^{--}$$

AsS_4^{3-} is the thio arsenate ion, and SnS_3^{--} the thio stannate ion.

Analysis of Copper Subgroup

3. The sulfides of the copper group (CuS, PbS, Bi_2S_3, CdS) are readily soluble in hot dil. HNO_3, for example,

$$3\ CuS + 8\ H^+ + 2\ NO_3^- \rightarrow 3\ Cu^{++} + 2\ NO + 2\ H_2O + 3\ S$$

4. When a solution of the cations (Cu^{++}, Pb^{++}, Bi^{3+}, Cd^{++}) has been obtained, lead may be precipitated as $PbSO_4$ by the addition of sulfate ion (H_2SO_4); sulfates of the three other cations are soluble, and therefore are not precipitated.

$$Pb^{++} + SO_4^{--} \rightarrow \underline{PbSO_4}$$

In the identification of lead, $PbSO_4$ is dissolved in $NH_4C_2H_3O_2$ solution, and Pb^{++} is precipitated as yellow $PbCrO_4$ on addition of K_2CrO_4. $PbSO_4$ dissolves in ammonium acetate because $Pb(C_2H_3O_2)_2$ is almost completely un-ionized.

$$PbSO_4 \rightleftharpoons \boxed{Pb^{++}} + SO_4^{--}$$
$$\boxed{2\ C_2H_3O_2^-}$$
$$\downarrow$$
$$Pb(C_2H_3O_2)_2$$

5. The basis for the separation of Bi^{3+} from Cd^{++} and Cu^{++} is the formation of insoluble $Bi(OH)_3$ on the addition of NH_4OH to a solution of the three ions.

$$Bi^{3+} + 3\ OH^- \rightarrow \underline{Bi(OH)_3}$$

While $Cu(OH)_2$ and $Cd(OH)_2$ are insoluble in water, both are soluble in excess NH_4OH solution because of the formation of complex ions.

$$Cu^{++} + 4\ NH_3 \rightarrow \underset{\text{blue}}{Cu(NH_3)_4^{++}}$$

$$Cd^{++} + 4\ NH_3 \rightarrow \underset{\text{colorless}}{Cd(NH_3)_4^{++}}$$

Except for some transition metals, a trivalent metal ion forms a hydroxide precipitate with NH_4OH, but a divalent metal ion forms a complex ion with NH_3.

Bismuth forms no complex ion with NH_3; hence the hydroxide remains insoluble in excess NH_4OH. Bismuth may be confirmed by the addition of sodium stannite solution to the precipitate of $Bi(OH)_3$. $Bi(OH)_3$ is instantly reduced to a finely divided black precipitate of Bi.

$$2\ Bi(OH)_3 + 3\ SnO_2^{--} \rightarrow 2\ \underline{Bi} + 3\ SnO_3^{--} + 3\ H_2O$$

6. The intense blue color of $Cu(NH_3)_4^{++}$ is evidence of the presence of Cu^{++}. Copper may be confirmed by precipitation of red-brown cupric ferrocyanide from an acid solution with ferrocyanide ion.

$$Cu(NH_3)_4^{++} + 4\ H^+ \rightarrow Cu^{++} + 4\ NH_4^+$$

$$2\ Cu^{++} + Fe(CN)_6^{4-} \rightarrow \underset{\text{red-brown}}{\underline{Cu_2Fe(CN)_6}}$$

7. Cadmium is identified by precipitation of CdS from a cyanide solution with H_2S. Cyanide ion forms a complex ion with Cd^{++}, which dissociates to give sufficient Cd^{++} for formation of CdS with H_2S.

$$Cd(NH_3)_4^{++} + 4\ CN^- + 4\ H_2O \rightarrow Cd(CN)_4^{--} + 4\ NH_4OH$$

$$Cd(CN)_4^{--} \leftrightharpoons Cd^{++} + 4\ CN^-$$

$$Cd^{++} + H_2S \rightarrow \underset{\text{yellow-orange}}{\underline{CdS}} + 2\ H^+$$

Copper ion is reduced in cyanide solution and remains as part of a monovalent cyanide complex ion of very slight ionization to copper ion. It is thus not precipitated by H_2S and does not interfere with yellow CdS precipitation.

$$2\ Cu(CN)_4^{--} \rightarrow 2\ Cu(CN)_2^- + (CN)_2 + 2\ CN^-$$

Analysis of the Arsenic Subgroup

When acid is added to the solution of complex thio salts of the arsenic group, the members of the group are reprecipitated as sulfides:

$$2\ AsS_4^{3-} + 6\ H^+ \rightarrow \underline{As_2S_5} + 3\ H_2S \uparrow$$

$$2\ SbS_4^{3-} + 6\ H^+ \rightarrow \underline{Sb_2S_5} + 3\ H_2S \uparrow$$

$$SnS_3^{--} + 2\ H^+ \rightarrow \underline{SnS_2} + H_2S \uparrow$$

$$HgS_2^{--} + 2\ H^+ \rightarrow \underline{HgS} + H_2S \uparrow$$

Note that all the members of this group are reprecipitated in their higher oxidation states.

8. The basis for the separation of Sb and Sn from Hg and As is the solubility of the sulfides of the former two elements in conc. HCl and the insolubility of HgS and As_2S_5 in that reagent. Treatment of the sulfides of the group with conc. HCl gives the reactions

$$SnS_2 + 4\ H^+ \rightarrow Sn^{4+} + 2\ H_2S$$

$$Sb_2S_5 + 6\ H^+ \rightarrow 2\ Sb^{3+} + 3\ H_2S + 2\ S$$

9. The resulting solution of $SnCl_4$ and $SbCl_3$ can be tested for Sb^{3+} without removing Sn^{4+} by adding oxalic acid and H_2S. Antimony is precipitated as the orange sulfide.

$$2\ Sb^{3+} + 3\ H_2S \rightarrow \underset{\text{orange-red}}{\underline{Sb_2S_3}} + 6\ H^+$$

Sn^{4+} forms a complex ion with oxalic acid and is not precipitated.

10. In testing for tin it is necessary to reduce stannic ion to stannous ion. This may be accomplished by the addition of iron (such as a small piece of piano wire or clean iron filings) or magnesium.

$$Sn^{4+} + Fe \rightarrow Sn^{++} + Fe^{++}$$

$$Sn^{4+} + Mg \rightarrow Sn^{++} + Mg^{++}$$

Antimony in the solution is simultaneously reduced to the free element which appears as a flaky black residue.

$$2\ Sb^{3+} + 3\ Fe \rightarrow 3\ Fe^{++} + 2\ \underline{Sb}$$

$$2\ Sb^{3+} + 3\ Mg \rightarrow 2\ \underline{Sb} + 3\ Mg^{++}$$

After obtaining a solution of tin in the stannous state, addition of $HgCl_2$ solution produces a grey precipitate composed of Hg (black) and Hg_2Cl_2 (white).

$$Hg^{++} + Sn^{++} \rightarrow \underline{Hg} + Sn^{4+} \text{ (excess } SnCl_2)$$

$$2\ Cl^- + 2\ Hg^{++} + Sn^{++} \rightarrow \underline{Hg_2Cl_2} + Sn^{4+} \text{ (excess } HgCl_2)$$

The relative amounts of Hg and Hg_2Cl_2 (and therefore the color of the precipitate) depend on the relative amounts of $SnCl_2$ and $HgCl_2$ present in the test.

11. The basis of the separation of As_2S_5 and HgS is the insolubility of HgS in NH_4OH. As_2S_5 dissolves according to the equation

$$As_2S_5 + 6\ NH_4OH \rightarrow (NH_4)_3AsO_3S + (NH_4)_3AsS_4 + 3\ H_2O$$

12. The test for arsenic consists of the reprecipitation of As_2S_5 from the ammoniacal solution with hydrochloric acid.

$$(NH_4)_3AsO_3S + (NH_4)_3AsS_4 + 6\ HCl \rightarrow \underset{\text{yellow}}{\underline{As_2S_5}} + 6\ NH_4Cl + 3\ H_2O$$

13. The test for mercury is made by dissolving HgS in aqua regia.

$$HgS + 4\ H^+ + NO_3^- + Cl^- \rightarrow Hg^{++} + NOCl + 2\ H_2O + S$$

Addition of $SnCl_2$ to the solution results in the precipitation of Hg and Hg_2Cl_2. Reactions are the same as given in the test for tin.

B. Preliminary tests for Group II. Use opposite page for notes and equations.

1. Colors of Sulfides
   Put 2 drops of t.s. of each ion of Group II (Hg^{++}, Pb^{++}, Bi^{3+}, Cd^{++}, Cu^{++}, ~~As^{3+}, Sb^{3+}~~, Sn^{++}) into as many separate test tubes. Dilute each to 0.5 ml, add 6 drops of thioacetamide solution, and heat in a boiling water bath for 5 minutes. After noting the colors of the ppts., cfg. and discard the supernatant liquids. Save the ppt. for the next part of this experiment. (Arsenic precipitates best in a hot HCl solution.)

2. Solubility of Sulfides
   To each of the above ppts. add 2 drops of sodium polysulfide, Na_2S_2, and 2 drops H_2O. Stir and warm. Record which dissolves rapidly, which slowly, and which not at all.

3. Methyl Violet Color Changes
   By dilution of 1 *M* HCl in a 10 ml graduate, prepare 0.5, 0.3, and 0.1 *M* solutions. Place some of each in as many test tubes, and place H_2O in a fourth. With a stirring rod add a drop of each to methyl violet papers and record the colors produced.

4. Hydrolysis of Bismuth and Antimony Salts
   To 1 drop of Bi^{3+} t.s. add water drop by drop until ppt. occurs. Will this dissolve in 1 *M* HCl? Repeat with Sb^{3+}.

5. Precipitation and Dissolving of Lead Sulfate
   To 2 drops Pb^{++} t.s. add 1 drop dil. H_2SO_4 and 5 drops of H_2O. Cfg. and discard liquid. Dissolve ppt. in a few drops of ammonium acetate solution. Warm if necessary. Add a drop of potassium chromate.

6. Reduction of Bismuth Hydroxide by Stannite Ion (SnO_2^{--})
   To 1 drop of Bi^{3+} t.s. in a test tube, add 1 drop of dil. NH_4OH. In another test tube place 2 drops of $SnCl_2$ and add NaOH to it drop by drop until the ppt. first formed dissolves. Now add 1 drop of this solution to the mixture in the other test tube.

7. Cd^{2+} is borderline between gr. II & III — use NH_4CN to adjust ph
   using 2 drops Cd^{2+}, adjust ph to .5 (= .3 m)
   then use thioacetamide to add H_2S. Observe.
   Trying to make CdS. If no ppt is formed, wrong ph was used.

C. Analytical procedures for Group II. The following procedures are to be used for analyzing known and unknown solutions. Prepare a known sample by mixing 2 drops of each test solution of ions of this group. (Use 4 drops each of Cd^{++}* and Pb^{++} test solutions; use Sn^{4+} rather than Sn^{++}, since the latter would reduce Hg^{++}.)

Analyze according to Procedures 5–7 below.

PROCEDURE 5 *Precipitation and Separation of Copper and Arsenic Subgroups*

SOLUTION (from Procedure 4, or sample known to contain Group II). Adjust the volume of solution to about 1 ml by evaporation or addition of H_2O (1). Neutralize the solution by adding conc. NH_4OH dropwise until just basic to litmus. Now adjust the acidity of the solution to 0.3 *M* by adding 1 *M* HCl dropwise until a blue-green color is obtained with methyl violet test paper (2). (The acidity may be adjusted according to Note 3 below.) Saturate the solution with H_2S.† Cfg. and, if the solution is not clear, add 2 drops NH_4Cl solution and 2 drops H_2O (4), (5). Heat to boiling and cfg. again.

PRECIPITATE: (6) HgS (black), CuS (black), Bi_2S_3 (brown-black), PbS (black), CdS (yellow), As_2S_3–As_2S_5 (yellow), SnS (brown-black), SnS_2 (yellow), Sb_2S_3–Sb_2S_5 (orange). Wash the ppt. with 10 drops of H_2O and discard the washings. To the ppt. add 3 drops Na_2S_2 and 7 drops H_2O. Warm and stir for 1–2 minutes. Cfg. Decant or draw off the solution. Repeat the extraction of the ppt. with Na_2S_2 as above and combine the two extraction solutions. Wash the ppt. once with 6 drops H_2O, combining the washing with the two extraction solutions.

SOLUTION. Boil the solution to remove H_2S. If there is a residue, cfg. and discard the ppt. Treat the solution for Groups III, IV, and V according to Procedure 8.

PRECIPITATE: Copper subgroup. Treat the ppt. according to Procedure 6.

SOLUTION: Arsenic subgroup. Treat according to Procedure 7.

† Thioacetamide may be used as a source of H_2S in any or all of the procedures. Whenever saturation with H_2S is called for, simply add 6 to 8 drops of 8% thioacetamide solution, place the test tube in a beaker of boiling water for 5 minutes, and proceed according to directions.

NOTES.

1. If the solution from the silver group ppt. is concentrated here, some SbOCl or BiOCl may ppt. Such a ppt., however, should not be separated, since these substances are sufficiently soluble so that they will be converted to sulfides with H_2S.

2. If the solution is less than 0.1 *M* acid, some ZnS may ppt. in this group; if the acidity is too high, PbS and CdS may fail to ppt. Methyl violet (m.v.) shows the following color changes:

Solution	*Color*
Neutral and alkaline	Violet
0.1 *M* H^+	Blue
0.3 *M* H^+	Blue-green
0.5 *M* H^+	Yellow-green
1.0 *M* H^+	Yellow

Cu^{2+} Bi^{3+} Cd^{2+}
Hg^{2+} Sn^{2+}
Pb^{2+} + one from above

* Since Cd^{++} is one of the more difficult to precipitate, it may be omitted at the option of the instructor.

3. The proper acidity may be obtained here without the use of methyl violet paper if, after the solution has been carefully neutralized with NH_4OH, *exactly* 2 drops of 6 *M* HCl are added.

4. Dilution of the sample reduces the H^+ concentration, which facilitates pptn. of the more soluble sulfides (CdS, SnS, PbS).

5. Addition of ammonium ion helps coagulate colloidal sulfides. Formation of colloidal sulfides can at times be troublesome. The use of a few granules of solid NH_4Cl to furnish ions to help coagulate the colloid, and patient heating with stirring usually renders the suspension susceptible to settling on centrifugation.

6. The color of the ppt. at this point may be indicative of ions present. A yellow ppt. would eliminate mercury, copper, bismuth, and lead, which form black sulfides.

PROCEDURE 6 *Analysis of Copper Subgroup*

PRECIPITATE (from Procedure 5 – CuS, PbS, Bi_2S_3, CdS, HgS). Add 6 drops 6 *M* HNO_3 and 7–10 drops H_2O (1). Heat to boiling and keep the solution hot for 1–2 minutes. Cfg. If any appreciable amount of ppt. remains, again treat with HNO_3 and combine the two solutions.

PRECIPITATE: S, HgS. Discard (2).	**SOLUTION:** Pb^{++}, Cu^{++}, Bi^{3+}, Cd^{++}. Transfer to a crucible or small casserole, add 3–4 drops conc. H_2SO_4 (3). Evaporate cautiously until dense white fumes appear. Cool. Add 10 drops H_2O (5). Stir thoroughly. Finally, transfer to a test tube and cfg. A white ppt. indicates lead. Decant off the liquid. Wash ppt. with 5 drops H_2O and 1 drop dil. H_2SO_4. Add the washings to the liquid above.
PRECIPITATE: $PbSO_4$ (white). Add 3 drops hot $NH_4C_2H_3O_2$ to the ppt. (4). Stir thoroughly. Add 1 drop K_2CrO_4. Formation of a yellow ppt. confirms *lead* (6).	**SOLUTION:** Bi^{3+}, Cu^{++}, Cd^{++}. Add 15 *M* NH_4OH dropwise until basic to litmus. A blue color indicates copper (7); a white flocculent ppt. indicates bismuth. Cfg.
PRECIPITATE: $Bi(OH)_3$ (white). Prepare Na_2SnO_2 by placing 2 drops $SnCl_2$ in test tube and adding NaOH until the ppt. formed dissolves (8); this may require 8–12 drops NaOH. Add this solution to ppt. above. Formation of a black ppt. within a few seconds confirms *bismuth*.	**SOLUTION:** $Cu(NH_3)_4^{++}$ (blue), $Cd(NH_3)_4^{++}$. Concentrate to about 1 ml. (*a*) To 4 drops of the solution add 6 *M* HCl until acid; then 2 drops $K_4Fe(CN)_6$. Cfg. A red ppt. [$Cu_2Fe(CN)_6$] confirms *copper*. (*b*) To the remainder of solution, add KCN dropwise until blue color disappears, then 2 drops in excess. (Add 1 drop KCN if Cu was absent.) Saturate with H_2S (by using thioacetamide). A yellow or orange ppt. proves *cadmium* (9).

NOTES.

1. The HNO_3 used to dissolve the sulfides of the copper group must be dilute, since conc. HNO_3 may oxidize PbS to $PbSO_4$.

$$PbS + 8\ HNO_3 \rightarrow PbSO_4 + 4\ H_2O + 8\ NO_2$$

$PbSO_4$ is insoluble and would be cfgd. off and thus lost from the solution.

2. HgS is insoluble in dil. HNO_3, and if any remains undissolved by Na_2S_2 in the separation of the copper and arsenic groups, it will appear as an insoluble residue in the copper group analysis.

3. The H_2SO_4 is evaporated to fumes of SO_3 to expel HNO_3 from the solution. When fumes of SO_3 appear, all HNO_3 has been expelled, and the H_2SO_4 has started decomposing.

$$H_2SO_4 \rightarrow H_2O + SO_3\uparrow$$

It is necessary to expel HNO_3, since (1) $PbSO_4$ is moderately soluble in HNO_3, and (2) later in the procedure when H_2S is used, HNO_3 would oxidize H_2S to free sulfur, which makes the analysis difficult.

4. $PbSO_4$ is soluble in hot ammonium acetate solution because slightly ionized $Pb(C_2H_3O_2)_2$ is formed. $PbCrO_4$ may then be precipitated from this solution by adding CrO_4^{--}. (This is a good example of the application of comparative solubility products and ionization constants.)

$$K_{sp} \text{ for } PbSO_4 = 2.3 \times 10^{-8}$$

$$K_{sp} \text{ for } PbCrO_4 = 1.8 \times 10^{-14}$$

The ionization of $Pb(C_2H_3O_2)_2$ gives a Pb^{++} concentration intermediate between $PbSO_4$ and $PbCrO_4$.

5. After the solution has been evaporated to fumes of SO_3, it is diluted to dissolve any sulfates of Bi, Cu, or Cd which may have precipitated in the concentrated acid solution.

6. Most of the Pb^{++} is removed in Procedure 4 as $PbCl_2$; consequently, only small amounts of Pb^{++} will be found in Procedure 6.

7. The test for Cu^{++} with NH_3 is very sensitive, the blue color being visible where as little as 1 part of the $Cu(NH_3)_4^{++}$ ion is present per 25,000 parts of H_2O.

8. Sodium stannite solution is prepared by adding an excess of NaOH to a solution of $SnCl_2$ (stannous chloride). The reaction takes place in two steps:

$$Sn^{++} + 2\ OH^- \rightarrow Sn(OH)_2$$

$$Sn(OH)_2 + 2\ OH^- \rightarrow Sn(OH)_4^{--} \quad \text{or} \quad SnO_2^{--} + 2\ H_2O$$

The stannite solution must be made up just before using, since it oxidizes in the air to form a stannate solution which does not react with $Bi(OH)_3$.

9. CdS may ppt. in a colloidal condition, and the solution will appear yellow or orange. Addition of 1 or 2 drops of conc. NH_4OH will cause pptn. of CdS. A very pale yellow is probably due to some decomposition products of thioacetamide.

Equations for Group II

Lead

Copper

Bismuth

Cadmium

Arsenic

Mercury

Antimony

Tin

PROCEDURE 7 *Analysis of Arsenic Subgroup*

<table>
<tr><td colspan="3">SOLUTION (from Procedure 5): HgS_2^{--}, AsS_4^{3-}, SbS_4^{3-}, SnS_3^{--}. Add 6 M HCl dropwise. Stir after each drop until acid to litmus (1). Cfg. and discard liquid (2).</td></tr>
<tr><td colspan="3">PRECIPITATE: As_2S_5, HgS, Sb_2S_5, SnS_2, S. Add 10 drops conc. 12 M HCl. Heat to nearly boiling with stirring. Cfg. Separate solution and treat ppt. with conc. HCl as above and combine the two solutions. Treat the combined solutions with H_2S (2 drops thioacetamide for 5 minutes in hot water bath) to reprecipitate any HgS which may have dissolved. Add any ppt. to that already taken out.</td></tr>
<tr><td colspan="2">PRECIPITATE: HgS, As_2S_5. Add 10 drops 6 M NH_4OH. Stir for about 1 minute, then cfg. Separate solution from ppt. Wash ppt. with 5 drops NH_4OH (3).</td><td rowspan="2">SOLUTION: Sb^{3+}, Sn^{4+}. Boil this solution gently until H_2S has been completely expelled.* Divide the solution into two parts, about 1/3 in part (a) and 2/3 in part (b).
(a) To the solution add about 0.1 g oxalic acid solid; then add 8–10 drops H_2O. Warm and saturate with H_2S (use thioacetamide). Orange ppt. proves antimony (5).
(b) Add a short strip of Mg ribbon and then a few drops 6 M HCl. After reaction ceases, cfg. if there is a residue, and add 1 drop $HgCl_2$ to the clear liquid. A white or grey ppt. proves tin (6).</td></tr>
<tr><td>PRECIPITATE: HgS, S. Treat with 5 drops conc. 12 M HCl and 2 drops 6 M HNO_3. Transfer to a crucible and evaporate nearly to dryness. Add 3 drops H_2O. Cfg. and filter if necessary to remove sulfur. Transfer solution to a test tube and add 2 drops $SnCl_2$. A white or grey ppt. proves mercury.</td><td>SOLUTION: AsS_4^{3-}. Acidify with 6 M HCl. Formation of yellow ppt. proves arsenic (4).</td></tr>
</table>

* Use a piece of filter paper which has been moistened with $Pb(NO_3)_2$ solution. Place the paper over the tube. No blackening will appear after H_2S is expelled.

NOTES.

1. The acid used in reprecipitation of the arsenic group should be dilute since Sb_2S_5 and SnS_2 are soluble in moderately concentrated acid.

2. Addition of acid to a solution of Na_2S_2 results in formation of a white milky ppt. of sulfur.

$$Na_2S_2 + 2\ HCl \rightarrow 2\ NaCl + H_2S \uparrow + \underline{S}$$

Hence a ppt. will always be obtained when the solution of sodium polysulfide is acidified, whether members of the arsenic group are present or not. Sulfides of the arsenic group are colored and coagulate more readily than the free sulfur and so appear as a colored residue in the milky solution.

3. Test wash solution for complete extraction of the arsenic by acidifying a drop of the solution in a watch glass. No yellow ppt. will appear if the extraction is complete.

4. Arsenic may be confirmed as follows: Tear a piece of filter paper such that it can be rested and crease-clamped upon a 6″ test tube. Place a small crystal of $AgNO_3$ on top of the paper. Place the solution to be tested for arsenic in the test tube. Add a few zinc granules (or some powdered zinc), then several drops of 6 *M* HCl to generate hydrogen gas. The latter forms AsH_3 with arsenic present. Arsine (AsH_3) gas rising in the tube and through the filter paper turns an $AgNO_3$ crystal yellow, then black.

Antimony forms stibine (SbH_3) if present, which turns the crystal black only. If sulfur is a contaminant, use conc. NaOH instead of HCl to generate hydrogen gas with zinc, since H_2S will turn $AgNO_3$ black. $AgNO_3$ crystals should not be handled with the fingers as black stains may be produced.

5. If a large amount of Sn^{4+} is present, some SnS_2 may precipitate here. However it is yellow to tan in color in contrast to red-orange Sb_2S_3.

6. A test for tin may be made on the original sample. To 3–4 drops of original sample, add 6 *M* HCl until distinctly acid. Add 2 or 3 short strips of Mg ribbon. After reaction ceases, filter off any residue. To the clear solution add 2 drops $HgCl_2$. A white precipitate indicates tin.

D. Analysis of unknown. When you have completed the analysis of the known solution of ions for this group, write equations for reactions of each ion on the opposite page. After you pass a quiz on the procedure and equations, the instructor will issue an unknown containing ions of this group. *Use not more than 0.5 ml of the sample for analysis.* As you analyze the sample, develop a diagrammatic scheme for the analysis in the space below.

Group III The Aluminum-Nickel-Iron Group

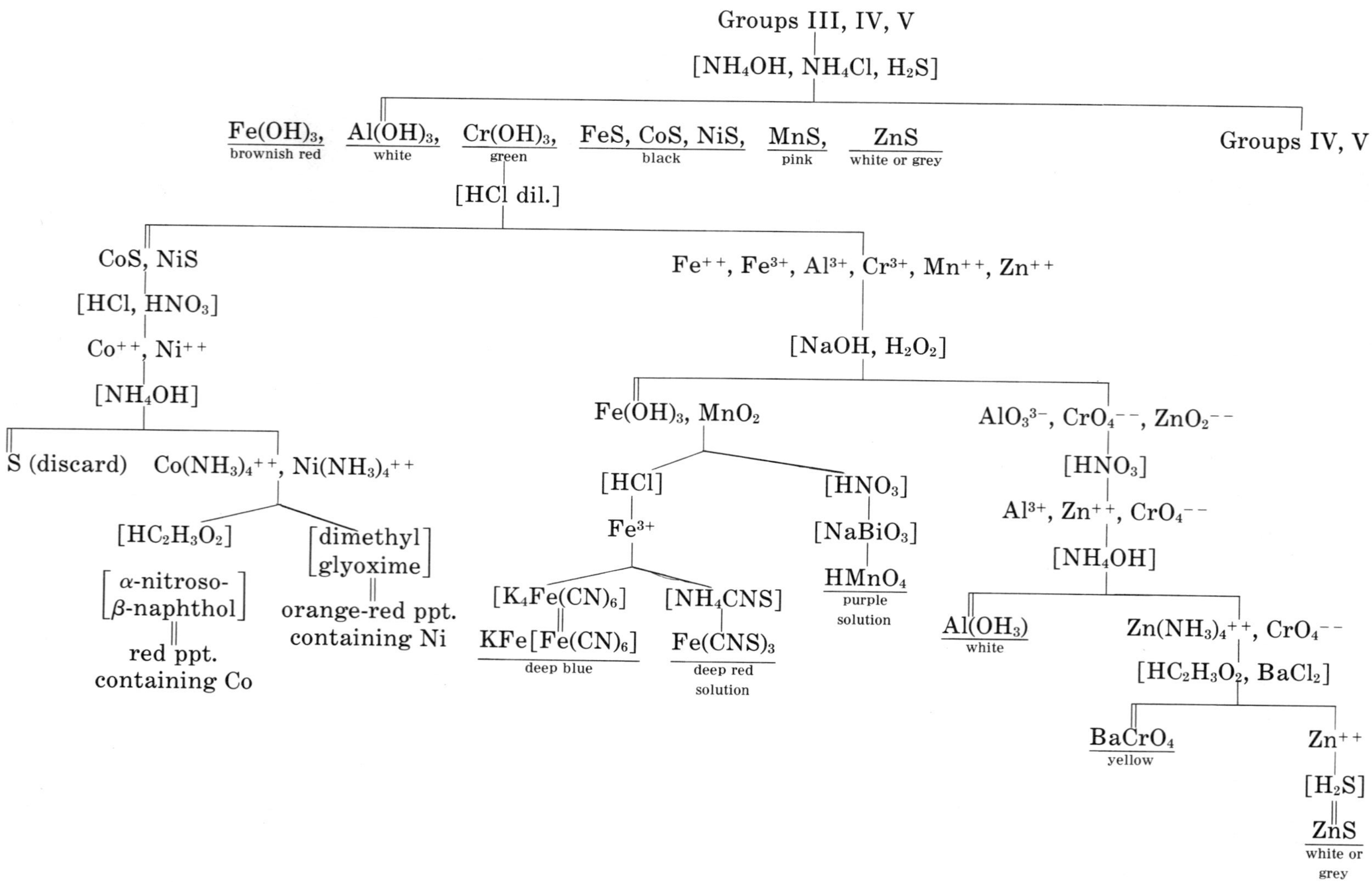

23
Group III
the aluminum-nickel-iron group

A. Chemistry of separations and tests.

Precipitation of Group III and Separation of Subgroups

1. The basis for the separation of Group III from Groups IV and V is the fact that sulfides and hydroxides of Group III precipitate in ammoniacal solution, while sulfides of Groups IV and V are soluble in this solution and hence do not precipitate. To prevent the precipitation of $Mg(OH)_2$, NH_4Cl is added to repress the ionization of NH_4OH and reduce the OH^- concentration to a point where $Mg(OH)_2$ cannot form.

   The separation of Group III from the moderately soluble hydroxides of Group IV and the slightly soluble $Mg(OH)_2$ of Group V depends on a carefully buffered solution. If NH_4OH is added to an aqueous solution of $MgCl_2$ and such a representative member of Group III as $AlCl_3$, the hydroxides of both magnesium and aluminum will precipitate. Therefore, it is necessary to reduce the hydroxide concentration to a point where $Al(OH)_3$ will precipitate but $Mg(OH)_2$ will not. Consider the following:

$$K_{sp} \text{ of } Mg(OH)_2 = 3.4 \times 10^{-11}$$

$$K_{sp} \text{ of } Al(OH)_3 = 2 \times 10^{-33}$$

   In 1 M NH_4OH, the concentration of OH^- is 4.24×10^{-3} g ion/liter. Then

$$[Mg^{++}] \times (4.24 \times 10^{-3})^2 = 3.4 \times 10^{-11}$$

$$[Mg^{++}] = 1.9 \times 10^{-6} \quad \text{(minimum conc. to start pptn.)}$$

$$[Al^{3+}] \times (4.24 \times 10^{-3})^3 = 2 \times 10^{-33}$$

$$[Al^{3+}] = 2.6 \times 10^{-25} \quad \text{(minimum conc. to start pptn.)}$$

   Now consider buffering the NH_4OH solution with NH_4Cl, so that the concentration of OH^- is decreased to 1.8×10^{-5}; then

$$[Mg^{++}] = \frac{3.4 \times 10^{-11}}{(1.8 \times 10^{-5})^2} = 0.11$$

   This concentration (0.11) of Mg^{++} is above that normally present in solutions for analysis; therefore, $Mg(OH)_2$ does not precipitate in Group III. However,

$Al(OH)_3$ and other hydroxides of Group III do precipitate. For example,

$$[Al^{3+}] = \frac{2 \times 10^{-33}}{(1.8 \times 10^{-5})^3} = 3.4 \times 10^{-19}$$

Above this concentration of Al^{3+}, $Al(OH)_3$ will precipitate – even in buffered NH_4OH.

2. Several of the ions in this group are colored, and often some information regarding the presence of cation members of Group III in the sample may be gained by observing the color of the solution. Colors of ions are: Co^{++} pink, Ni^{++} green, Fe^{++} very pale green, Cr^{3+} dark green, Mn^{++} pale pink to colorless. Fe^{3+} in the presence of Cl^- forms a complex ion which is yellow. Obviously, if the solution to be analyzed for Group III appears colorless, those ions with distinct colors (Ni^{++}, Cr^{3+}, Co^{++}) could be eliminated as possibilities.

3. Additional information can be obtained after the solution is made ammoniacal with NH_4OH. Zinc, nickel, and cobalt form complex ions and thus are not precipitated until after addition of H_2S. However, iron, chromium, and aluminum precipitate as hydroxides: $Fe(OH)_3$ red-brown, $Cr(OH)_3$ green, $Al(OH)_3$ white. If no precipitate formed on making the solution ammoniacal, then Fe^{3+}, Cr^{3+}, Al^{3+} could be eliminated as possibilities. Reactions of the ions with NH_4OH are

$$Ni^{++} + 4\ NH_3 \rightarrow Ni(NH_3)_4^{++}$$

$$Co^{++} + 4\ NH_3 \rightarrow Co(NH_3)_4^{++}$$

$$Zn^{++} + 4\ NH_3 \rightarrow Zn(NH_3)_4^{++}$$

$$Fe^{3+} + 3\ NH_4OH \rightarrow \underline{Fe(OH)_3} + 3\ NH_4^+$$

$$Al^{3+} + 3\ NH_4OH \rightarrow \underline{Al(OH)_3} + 3\ NH_4^+$$

$$Cr^{3+} + 3\ NH_4OH \rightarrow \underline{Cr(OH)_3} + 3\ NH_4^+$$

Just as trivalent Bi^{3+} precipitates as the hydroxide with NH_4OH and divalent Cu^{++} and Cd^{++} form complex ammoniates in Group II, so the various trivalent and divalent metals of Group III follow the same pattern.

4. On addition of H_2S, the three insoluble hydroxides above remain as such, but the other ions produce insoluble sulfides.

$$Ni(NH_3)_4^{++} + S^{--} \rightarrow \underline{NiS} + 4\ NH_3$$

$$Fe^{++} + S^{--} \rightarrow \underline{FeS}$$

Some $Fe(OH)_3$ may be converted to FeS.

$$2\ Fe(OH)_3 + 3\ H_2S \rightarrow \underline{2\ FeS} + \underline{S} + 6\ H_2O$$

5. The basis for the separation of Co and Ni from other members of the group depends on the fact that NiS and CoS are very slowly soluble in dil. HCl, while the sulfides and hydroxides of the other metals dissolve readily. In effecting

the separation it is very important, then, that NiS and CoS not remain in contact with HCl very long.

6. The separation of iron and manganese from aluminum, chromium, and zinc is based on the fact that the hydroxides of the latter three elements are amphoteric; that is, their hydroxides are soluble in excess strong base, such as NaOH.

$$\underline{Zn(OH)_2} + 2\ OH^- \rightarrow Zn(OH)_4^{--} \quad \text{or} \quad ZnO_2^{--} + 2\ H_2O$$

$$\underline{Al(OH)_3} + 3\ OH^- \rightarrow Al(OH)_6^{3-} \quad \text{or} \quad AlO_3^{3-} + 3\ H_2O$$

$$\underline{Al(OH)_3} + OH^- \rightarrow Al(OH)_4^{-} \quad \text{or} \quad AlO_2^{-} + 2\ H_2O$$

$$\underline{Cr(OH)_3} + 3\ OH^- \rightarrow Cr(OH)_6^{3-} \quad \text{or} \quad CrO_3^{3-} + 3\ H_2O$$

Hydrogen peroxide is added to convert $Cr(OH)_6^{3-}$ to CrO_4^{--}, so that aluminum can be effectively separated from chromium.

$$2\ Cr(OH)_6^{3-} + 3\ H_2O_2 \rightarrow 2\ CrO_4^{--} + 8\ H_2O + 2\ OH^-$$

$Mn(OH)_2$ and $Fe(OH)_2$ are also oxidized by H_2O_2.

$$\underline{Mn(OH)_2} + H_2O_2 \rightarrow \underline{MnO_2} + 2\ H_2O$$

$$\underline{2\ Fe(OH)_2} + H_2O_2 \rightarrow \underline{2\ Fe(OH)_3}$$

The above separations yield three subgroups: nickel subgroup (Co, Ni); iron subgroup (Fe, Mn); aluminum subgroup (Al, Cr, Zn).

Analysis of Nickel Subgroup

7. CoS and NiS are most readily dissolved by aqua regia.

$$CoS + 4\ H^+ + Cl^- + NO_3^- \rightarrow Co^{++} + S + NOCl + 2\ H_2O$$

Tests for Co and Ni may be made in the presence of one another; consequently, it is unnecessary to effect a separation. Organic reagents are used in testing for both of these ions; dimethyl glyoxime produces an orange-red precipitate with Ni^{++} in ammoniacal solution, and α-nitroso-β-naphthol produces a dark red precipitate with Co^{++} in acetic acid solution.

Analysis of Iron Subgroup

8. Iron and manganese ions may be tested for in the presence of each other by the following: Fe^{3+} is readily tested for by addition of $K_4Fe(CN)_6$, which gives a dark blue precipitate of $KFe[Fe(CN)_6]$, called *Prussian blue*

$$Fe^{3+} + K^+ + Fe(CN)_6^{4-} \rightarrow \underline{KFe[Fe(CN)_6]}$$

An alternate test is addition of CNS^- (NH_4CNS), which forms a red thiocyanate complex with Fe^{3+}, possibly $FeCNS^{++}$, $Fe(CNS)_2^+$, or $Fe(CNS)_3$.

9. The test for Mn is based on the oxidation of MnO_2 to violet-colored perman-

ganate ion, MnO_4^-. This is readily accomplished with PbO_2 or $NaBiO_3$ (sodium bismuthate) and HNO_3. HCl or Cl^- should be absent.

$$2\ MnO_2 + 3\ PbO_2 + 4\ H^+ \rightarrow 2\ MnO_4^- + 3\ Pb^{++} + 2\ H_2O$$

$$2\ MnO_2 + 3\ NaBiO_3 + 10\ H^+ \rightarrow 2\ MnO_4^- + 3\ Na^+ + 3\ Bi^{3+} + 5\ H_2O$$

10. The basis for the separation of Al^{3+} from CrO_4^{--} and Zn^{++} is the fact that NH_4OH reacts with Al^{3+} to form $Al(OH)_3$, which precipitates but forms a complex ion with Zn^{++}, namely $Zn(NH_3)_4^{++}$, which remains in solution. CrO_4^{--} is unaffected by NH_4OH.

$$Al^{3+} + 3\ NH_4OH \rightarrow \underline{Al(OH)_3} + 3\ NH_4^+$$

$$Zn^{++} + 4\ NH_3 \rightarrow Zn(NH_3)_4^{++}$$

The white flocculent precipitate of $Al(OH)_3$ on addition of NH_4OH indicates Al^{3+}. Very concentrated NH_4OH will redissolve $Al(OH)_3$.

$$Al(OH)_3 + 3\ OH^- \rightarrow Al(OH)_6^{3-}$$

Certain organic substances are adsorbed by $Al(OH)_3$ to give a colored product, and these are sometimes helpful in confirming its presence. Aurin tricarboxylic acid, called *aluminon*, gives a red color to the $Al(OH)_3$; similarly, arsenazo agent gives a purple color.

11. In separating Zn^{++} and CrO_4^{--}, advantage is taken of the fact that Cr is a part of the anion CrO_4^{--} and may be precipitated as $BaCrO_4$ in acetic acid solution on addition of Ba^{++}.

$$CrO_4^{--} + Ba^{++} \rightarrow \underline{BaCrO_4}$$

Zn^{++} then may be precipitated in the weakly acid filtrate.

$$Zn^{++} + H_2S \rightarrow \underline{ZnS} + 2\ H^+$$

Zinc may be confirmed by dissolving the ZnS precipitate in HCl, heating to remove H_2S, making the solution basic with NaOH, and adding a drop to dithizone test paper; a red-purple color confirms Zn^{++}. An alternate test is the following:

$$3\ Zn^{++} + 2\ K^+ + 2\ Fe(CN)_6^{4-} \rightarrow Zn_3K_2[Fe(CN)_6]_2$$

Zinc may be confirmed by treating the ZnS precipitate with HNO_3 and $Co(NO_3)_2$ and heating the mixture to a high temperature in a crucible; if the product is green, zinc may be assumed present.

$$CoO + ZnO \rightarrow CoZnO_2 \text{ (Rinmann's green)}$$

B. Preliminary tests on Group III. Use the opposite page for notes and equations.

1. Effect of NH_4OH and NH_4Cl on Group III Cations.
   Place 1 drop of t.s. on each ion of Group III (Fe^{3+}, Al^{3+}, Mn^{++}, Cr^{3+}, Ni^{++}, Co^{++}, Zn^{++}) into separate test tubes. Dilute each to about 0.5 ml, and add 1 drop of dil. NH_4OH to each. Write the equations. Now add 3 drops of conc. NH_4OH to each tube. Note which precipitates, if any, dissolve. Repeat the tests on Ni^{++}, Co^{++}, and Zn^{++}, first adding 2 drops of NH_4Cl solution.

2. Action of NaOH and H_2O_2 on $Cr(OH)_3$.
   Place 1 drop of Cr^{3+} t.s. in test tube and dilute to 0.5 ml. Add 1 drop NaOH. Cfg. the tube containing $Cr(OH)_3$, and discard the solution. To the ppt. add 2 drops of NaOH and a few drops of H_2O_2, and warm.

3. Effect of NaOH on Aluminum and Zinc Ions.
   To several drops of t.s. of Al^{3+} and Zn^{++} in separate test tubes add 1 drop of NaOH and observe. Then add an excess of NaOH to each.

4. A Means of Preventing $Mg(OH)_2$ from Precipating in Group III.
   To each of 2 test tubes add 1 drop of Mg^{++} t.s. (Mg is in Group V). To one tube add a drop of NH_4Cl solution. Now add NH_4OH to each tube. (See theoretical discussion on page 117.)

NH_4Cl + NH_3

5 Fe^{2+} Ni^{2+} Co^{2+} Mn^{2+} Zn^{2+}
separately, make buffer, add H_2S and Heat
Sulfides will be formed. Observe ppt careful
attn to color

C. Analytical procedures for Group III.

Make up a known solution using 3 drops of test solution of each ion of this group. Analyze according to Procedures 8–11.

PROCEDURE 8 *Precipitation and Separation of Subgroups of Group III*

SOLUTION (from Procedure 5, containing cations of Groups III, IV, V). (1) Adjust the volume of solution to about 1.5 ml; add 2 drops NH_4Cl solution (2); then add 6 *M* NH_4OH until just basic. Add 2 drops 15 *M* NH_4OH; then saturate with H_2S (3),(4), using thioacetamide. Cfg.

HEAT

PRECIPITATE: CoS (black), NiS (black), FeS (black), $Al(OH)_3$ (white), $Cr(OH)_3$ (green), MnS (pink), ZnS (white), $Fe(OH)_3$ (red-brown). Wash twice with H_2O. Reject washings. To the ppt. in test tube add 12 drops H_2O and 2 drops conc. 12 *M* HCl. Stir thoroughly for about 1 minute. Cfg. immediately.	**SOLUTION:** Add 6 *M* acetic acid until acid; evaporate to 1 ml (5). Cfg. and reject any ppt. Decant supernatant liquid and test for Groups IV and V by Procedure 12.

PRECIPITATE: CoS, NiS. Treat according to Procedure 9.	**SOLUTION:** Fe^{++}, Fe^{3+}, Mn^{++}, Cr^{3+} (green), Al^{3+}, Zn^{++}. Transfer to a crucible or small casserole; add NaOH until basic, then 4 drops more. Add 6–8 drops of 3% H_2O_2 slowly (6). Heat solution to near boiling for a few minutes and until the gas (oxygen) evolution ceases. Add 3 ml H_2O. Transfer to test tube and cfg.

PRECIPITATE: Iron subgroup: $Fe(OH)_3$ (brownish red), MnO_2 (brown to black). Wash ppt. twice with hot H_2O. Discard washings. Treat ppt. according to Procedure 10.	**SOLUTION:** Aluminum subgroup: AlO_3^{3-}, CrO_4^{--} (yellow), ZnO_2^{--}. Treat according to Procedure 11.

NOTES.

1. H_2S is removed before addition of NH_4OH to gain information about pptn. with NH_4OH alone. (See Procedure 5.) Any Group II sulfides which might be in filtrate in a colloidal form are pptd. when H_2S is expelled. These may be cfgd. and discarded.

2. NH_4Cl is added to prevent pptn. of $Mg(OH)_2$ and to promote coagulation of sulfides which tend to ppt. in colloidal form.

3. When H_2S is added to an ammoniacal solution, a high concentration of S^{--} results. S^{--} thus may react directly with cations to produce insoluble sulfides.

4. It is important that the solution not be too ammoniacal when H_2S is added. Excess NH_4OH tends to hold certain sulfides, notably CoS and NiS, in colloidal condition.

5. The filtrate of Groups IV and V is acidified with $HC_2H_3O_2$ and evaporated for removal of colloidal constituents which may have remained in suspension.

6. Sodium peroxide may be used here instead of H_2O_2. The former is added slowly from the tip of a spatula. (Any unused Na_2O_2 should be disposed of in the sink, because combustible material such as paper may be ignited because of the strong oxidizing character of the peroxide. Also, on exposure to air, Na_2O_2 becomes sticky and hard from the formation of NaOH and Na_2CO_3 from moisture and CO_2 in the air.)

4 known mixtures?

(1) Co^{2+} Ni^{2+}

(2) Cr^{3+} Al^{3+} Cr masks Al. Try to prove Al.
Cr behaves sim to Al, except Cr is colored

(3) Fe^{2+} Mn^{2+} Zn^{2+}

(4) Fe^{2+} Ni^{2+} mask
should do 4 flow charts or write ups

PROCEDURE 9 *Analysis of Nickel Subgroup*

<table>
<tr><td colspan="3">PRECIPITATE (from Procedure 8). Wash twice with cold H_2O and reject the washings. To the ppt. add 5 drops 6 M HCl and 1 drop 6 M HNO_3. Heat until ppt. dissolves. Add 5–6 drops H_2O. Add 15 M NH_4OH dropwise to the solution until basic, then 1 drop more. Cfg.</td></tr>
<tr><td rowspan="2">PRECIPITATE: Some S from oxidation of H_2S. Discard.</td><td colspan="2">SOLUTION: $Co(NH_3)_4^{++}$, $Ni(NH_3)_4^{++}$. Divide the filtrate into two equal parts, (a) and (b).</td></tr>
<tr><td>(a) Add acetic acid until acid, then 4 drops α-nitroso-β-naphthol. Warm and allow to stand 10–15 minutes (1),(2). Formation of a red ppt. proves cobalt.</td><td>(b) Add 3 drops dimethyl glyoxime and 1 drop 15 M NH_4OH. Red ppt. proves nickel.</td></tr>
</table>

NOTES.

1. The ppt. of cobalt with α-nitroso-β-naphthol forms very slowly. Hence, the solution should be allowed to stand 10–15 minutes before any conclusion is drawn about the absence of Co^{++}.

2. The α-nitroso-β-naphthol reagent must be freshly prepared to be satisfactory as a test for Co^{++} because it deteriorates rapidly. Before making the test on the unknown, it is well to use the reagent with a known sample of Co^{++}. If the test on the known is unsatisfactory, obtain fresh solution.

PROCEDURE 10 *Analysis of Iron Subgroup*

<table>
<tr><td colspan="2">PRECIPITATE (from Procedure 8). Divide ppt. into two parts; about 2/3 in part (a) and 1/3 in part (b).</td></tr>
<tr><td>(a) To part of the ppt. add 5 drops 6 M HNO_3 and 5 drops H_2O. Add a pinch of $NaBiO_3$ and mix thoroughly. Heat and cfg. A violet supernatant liquid proves manganese. (PbO_2 may be used in place of $NaBiO_3$. Cl^- should not be present in making this test.)</td><td>(b) Add 5 drops 6 M HCl and heat until dissolved. Divide into two parts. Cool and add 1 drop $K_4Fe(CN)_6$ (1),(2) to one part. Formation of a deep blue ppt. of Prussian blue proves iron. To the second portion add 1 drop NH_4CNS. A deep red solution verifies iron.</td></tr>
</table>

NOTES.

1. Whether originally present as Fe^{++} or Fe^{3+}, iron will appear at this point in the procedure as Fe^{3+}. To determine the state of oxidation of iron in the original sample, solutions of potassium ferrocyanide and potassium ferricyanide may be used. Dissolve a small portion of the original sample in dil. HCl. Divide into two parts; to the first part add a few drops of $K_4Fe(CN)_6$. A deep Prussian blue ppt. proves ferric ion. To the second part add a few drops of $K_3Fe(CN)_6$. A ppt. of deep Prussian blue proves presence of ferrous ion. The same blue precipitate is obtained in both tests.

$$Fe^{3+} + K^+ + Fe(CN)_6^{4-} \rightarrow \underset{\text{Prussian blue}}{\underline{KFe[Fe(CN)_6]}}$$

$$Fe^{++} + K^+ + Fe(CN)_6^{3-} \rightarrow \underset{\text{Prussian blue}}{\underline{KFe[Fe(CN)_6]}}$$

If a deep blue ppt. is obtained with both ferro- and ferricyanide solutions, then both ferrous and ferric ions are present. A solution containing Fe^{3+} forms a deep red solution of $FeCNS^{++}$, $Fe(CNS)_2^+$, or $Fe(CNS)_3$ with CNS^-.

2. A light blue color in the test for iron is probably due to a trace of iron or some other metallic ion which has not been removed completely. If the test does not give a *deep, dark blue* ppt., iron may be assumed to be absent.

PROCEDURE 11 *Analysis of Aluminum Subgroup*

SOLUTION (from Procedure 8). Add 6 *M* HCl in a small evaporating dish until just acid (1); then add 15 *M* NH_4OH dropwise until basic. Add 3 more drops NH_4OH. Heat to boiling. Transfer to test tube. Formation of white flocculent ppt. proves *aluminum*. Cfg.

PRECIPITATE: $Al(OH)_3$ (white) (7).

SOLUTION: CrO_4^{--} (yellow), $Zn(NH_3)_4^{++}$. If the solution is yellow, *chromium* is present (2). Acidify with acetic acid (3). Add a few crystals of $NaC_2H_3O_2$, and then 2 drops $BaCl_2$. Formation of a yellow ppt. confirms *chromium*. Cfg.

PRECIPITATE: $BaCrO_4$ (yellow) and $BaSO_4$ (white).

SOLUTION: Divide solution into two parts. Add 3–4 drops of 0.1 *M* $K_4Fe(CN)_6$ to one part for slightly grey $Zn_3K_2[Fe(CN)_6]_2$ ppt. Saturate other portion with H_2S using thioacetamide. A grey-white ppt. is probably ZnS (4). To confirm, cfg. and wash ppt. twice with H_2O. Reject washings. Add 6–8 drops 6 *M* HCl to residue. If ppt. dissolves, presence of *zinc* is confirmed (5),(6).

NOTES.

1. Chromium chemistry involves several oxidation states. Cr^{3+} treated with excess NaOH forms CrO_3^{3-} [$Cr(OH)_6^{3-}$, chromite], much as Al^{3+} treated with excess NaOH forms AlO_3^{3-} [$Al(OH)_6^{3-}$, aluminate]. Addition of acid (HCl or HNO_3) converts back to Cr^{3+} and Al^{3+}. If CrO_3^{3-} has not been oxidized to yellow stable CrO_4^{--}, Cr^{3+} will follow Al^{3+} and a mixed ppt. of $Al(OH)_3$ and $Cr(OH)_3$ forms. The $Cr(OH)_3$ would thus interfere with the final $Al(OH)_3$ test. $Cr(OH)_3$ in small amounts is light green and may be mistaken for $Al(OH)_3$.

The reason for 6 *M* HCl instead of 6 *M* HNO_3 in the treatment of AlO_3^{3-}, ZnO_2^{--}, CrO_4^{--} to convert to Al^{3+}, Zn^{++}, CrO_4^{--} is because CrO_4^{--} in nitrate-containing solution and with residual H_2O_2 may yield an indigo-blue peroxide compound to which the formula CrO_5 has been given, and which compound may decompose to Cr^{3+}.

$$2\ CrO_4^{--} + 2\ H^+ \rightleftharpoons Cr_2O_7^{--} + H_2O$$

$$2\ H^+ + Cr_2O_7^{--} + 4\ H_2O_2 \rightleftharpoons 2\ CrO_5 + 5\ H_2O$$

The unstable colored CrO_5 more or less rapidly decomposes with oxygen evolution (decomposition rate is dependent on HNO_3 and H_2O_2 concentration) to Cr^{3+} ion.

$$4\ CrO_5 + 12\ H^+ \rightarrow 4\ Cr^{3+} + 6\ H_2O + 7\ O_2$$

The indigo-blue CrO_5 is soluble and more stable in ether. Thus a verifying test for yellow CrO_4^{--} is to use a $BaCrO_4$ ppt. suspended in 2 ml H_2O or 2 ml yellow CrO_4^{--} solution in a test tube. Add 5 drops 3 *M* HNO_3 and heat to boiling. Then cool under tap water. Add 1 ml ethyl ether to float above the CrO_4^{--} solution and 2 drops 3% H_2O_2. Stir. A blue color in the ether layer confirms chromium.

2. If CrO_4^{--} is present, the filtrate from Procedure 8 should be yellow. The yellow color is a sensitive test for CrO_4^{--}, and if solution from Procedure 8 is colorless, assume that chromium is absent and go directly to the test for zinc.

3. Since $BaCrO_4$ is soluble in HCl, HNO_3, and other strong acids, the pptn. of $BaCrO_4$ is made in $HC_2H_3O_2$ in which it is insoluble. The use of $NaC_2H_3O_2$ is to yield ions to help ppt. $BaCrO_4$ which tends to be colloidal. A few minutes heating in a water bath should yield $BaCrO_4$ ppt. See also Note 1 for addition confirmatory test for chromium.

4. In pptn. of ZnS with thioacetamide, some sulfur may ppt. in colloidal form; this white ppt. may be mistaken for white ZnS. Both ZnS and Zn_3K_2 $[Fe(CN)_6]_2$ are insoluble in slightly acid solution. Here the solution from the above manipulation should be slightly acid with acetic acid.

5. Zinc may also be confirmed by adding excess 5% NaOH to a few drops of the original sample and testing with dithizone test paper. A red-violet color to the test paper is produced if Zn^{++} is present. NaOH gives an orange color which must not be mistaken for the red-violet color with Zn^{++}. A known sample of Zn^{++} should always be used side by side with the unknown when using the dithizone test paper.

6. Zinc may be confirmed by making a paste of pptd. ZnS with 2 drops 0.05 *M* $Co(NO_3)_2$ and 1 drop dil. HNO_3. [A very small crystal of $Co(NO_3)_2$ may be used instead of 2 drops.] Place paste in a small crucible (or evapor-

ating dish). Dry carefully and heat the crucible bottom to redness in the hottest part of the Bunsen flame (high temperature essential). Formation of green $CoZnO_2$ confirms zinc. Colored glaze on porcelain is made commercially in this manner. Here, the green glaze can be wiped off.

7. To confirm Al^{3+}, suspend ppt. in 10 drops H_2O; then add dil. acetic acid until solution is slightly acid. Add 4 drops aluminon reagent, and heat in water bath. When it is heated add a few drops of dil. NH_4OH until it is neutral or slightly basic. Red ppt. confirms Al^{3+}. In all tests using organic reagents, run a known sample along with the unknown for a comparison. The arsenazo reagent test for Al is made as follows: Dissolve some $Al(OH)_3$ ppt. in dil. $HC_2H_3O_2$ (pH = 3–7). Add 2 drops arsenazo reagent (0.001 *M*). If Al is present, a reddish purple color results; if Al is absent, an orange color results.

Equations for Group III

Cobalt

Nickel

Iron

Manganese

Aluminum

Chromium

Zinc

D. Analysis of unknown solution. When the analysis of the known solution has been completed, write equations for all reactions of ions of this group on the opposite page. The instructor will then quiz you on the procedures and reactions of this group. After you satisfactorily pass the quiz, you will receive an unknown of ions in this group. Use 0.5 ml of the sample for analysis. Analyze according to Procedures 8–11, developing a diagrammatic scheme of analysis as you progress. Use the space below for the diagrammatic scheme.

Group IV The Barium Group

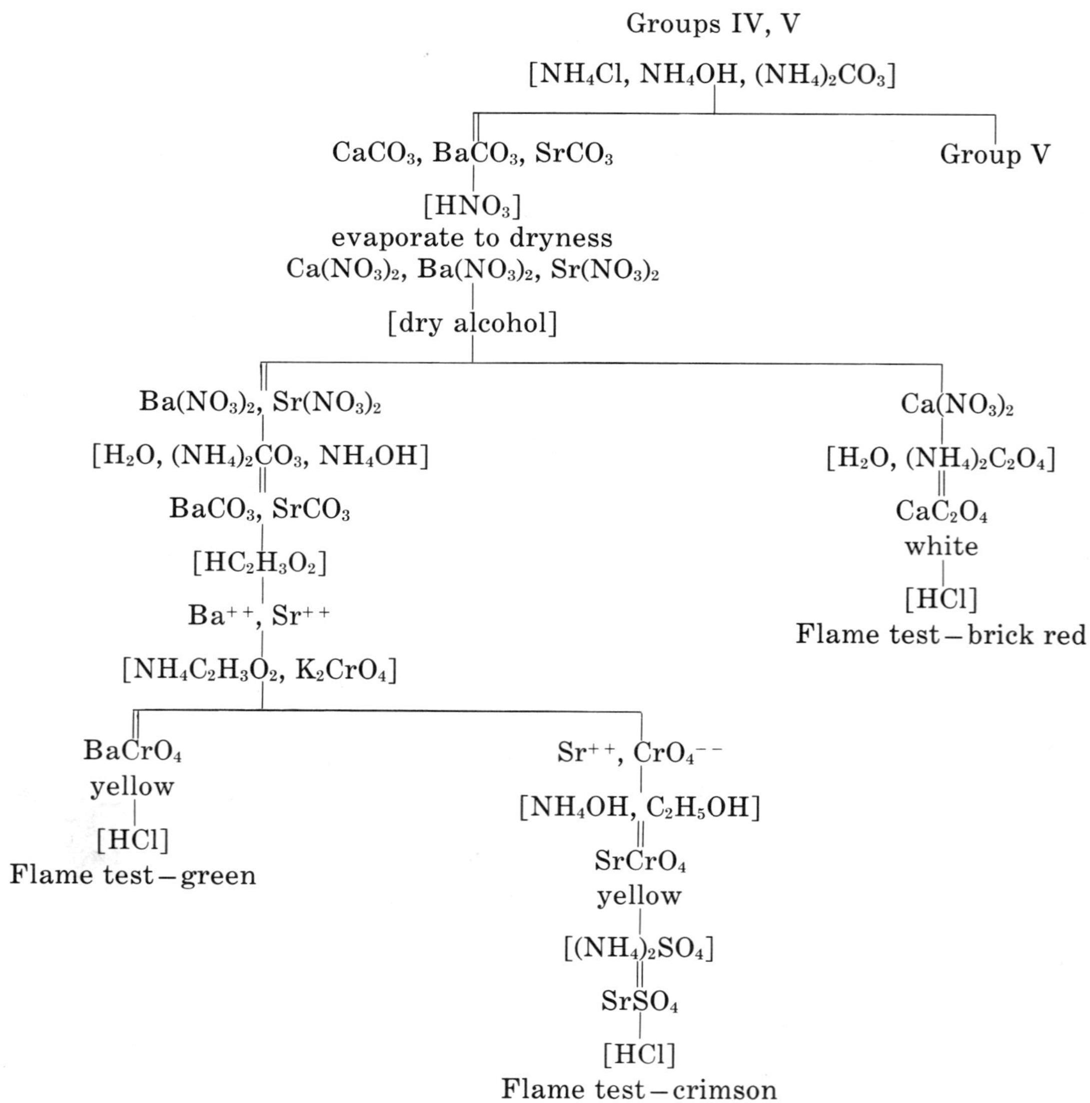

24
Group IV
the barium group

A. Chemistry of separations and tests.

SEPARATION OF BARIUM AND ALKALI GROUPS AND ANALYSIS OF THE BARIUM GROUP

1. The addition of $(NH_4)_2CO_3$ to an alkaline solution of the Ba and alkali groups in the presence of excess $NH_4^+(NH_4Cl)$ results in the precipitation of $BaCO_3$, $SrCO_3$, and $CaCO_3$. Mg^{++}, K^+, Na^+, NH_4^+ remain in solution.

$$Ba^{++} + CO_3^{--} \rightarrow \underline{BaCO_3}$$

$$Sr^{++} + CO_3^{--} \rightarrow \underline{SrCO_3}$$

$$Ca^{++} + CO_3^{--} \rightarrow \underline{CaCO_3}$$

2. The separation of Ba^{++} and Sr^{++} from Ca^{++} is based on the slight solubility of the nitrates of barium and strontium in dry (100%) alcohol; $Ca(NO_3)_2$ is soluble. For example, the solubilities at 25° C in mg of salt per ml of dry ethyl alcohol are approximately

$Ca(NO_3)_2 - 850$ $\quad$ $Ba(NO_3)_2 - 0.02$ $\quad$ $Sr(NO_3)_2 - 0.06$

Dry isopropyl alcohol or acetone may be used in place of ethyl alcohol.

3. The separation of Ba^{++} from Sr^{++} is based on the fact that $BaCrO_4$ is insoluble in acetic acid solution, whereas $SrCrO_4$ is soluble. Thus if chromate ion (K_2CrO_4) is added to an acetic acid solution of Ba^{++} and Sr^{++}, only the Ba^{++} is precipitated as $BaCrO_4$.

$$Ba^{++} + CrO_4^{--} \rightarrow \underline{BaCrO_4}$$

On addition of alcohol and OH^- to the solution, the solubility of $SrCrO_4$ is decreased, and it precipitates.

4. Ca^{++} is precipitated with oxalate ion, $C_2O_4^{--}$ in ammoniacal solution:

$$Ca^{++} + C_2O_4^{--} \rightarrow \underline{CaC_2O_4}$$

B. Preliminary tests for Group IV. There is no necessity for preliminary tests for this group. Go at once to the analytical procedures.

Equations for Group IV

Barium

Strontium

Calcium

C. Analytical procedures for Group IV. Make up a known sample of ions of this group by using 4 drops each of test solutions of the ions. Analyze according to Procedure 12. Write equations for reactions of the ions of the opposite page.

PROCEDURE 12 *Separation of Barium and Alkali Groups and Analysis of Barium Group*

SOLUTION (from Procedure 8). If NH_4^+ is not already present, add 2 drops NH_4Cl (1). Add 6 *M* NH_4OH dropwise until the solution is basic. Add 5 drops $(NH_4)_2CO_3$ (2), and place the tube in a hot water bath for a few minutes. Stir occasionally. Cfg. Add 1 drop more of $(NH_4)_2CO_3$ to the supernatant liquid to test for complete pptn. Decant the solution.

PRECIPITATE: $BaCO_3$, $SrCO_3$, $CaCO_3$ (all white). Wash ppt. with a few drops of H_2O to which 1 drop $(NH_4)_2CO_3$ has been added. Reject the washings. Dissolve ppt. using 8 drops H_2O and 2 drops 6 *M* HNO_3. Evaporate to dryness in an evaporating dish, but do not bake. Stir and digest the residue with 2 ml absolute ethyl alcohol or isopropyl alcohol. Add more alcohol if necessary to keep the volume about 2 ml. Transfer solution and residue to small test tube. Cfg. and decant the solution (3).

SOLUTION: Group V ions. Treat according to Procedure 13.

RESIDUE: $Ba(NO_3)_2$, $Sr(NO_3)_2$. Add 1 ml H_2O to dissolve, then reprecipitate $BaCO_3$, $SrCO_3$ with $(NH_4)_2CO_3$ as above. Cfg. Discard liquid. Dissolve ppt. in 3–4 drops 6 *M* acetic acid. Add 2 drops $NH_4C_2H_3O_2$ and 5 drops H_2O. Add 2 drops K_2CrO_4 (4). Stir for about 1 minute. A yellow ppt. indicates *barium*. Cfg.

SOLUTION: $Ca(NO_3)_2(Mg^{++})$. Add 2 ml H_2O. Evaporate to 1 ml and discard any ppt. Add 2 drops dil. NH_4OH and 2 drops $(NH_4)_2C_2O_4$. A white ppt. indicates calcium. Warm 2 minutes. Cfg. Save solution for Mg^{++} test below. Treat the ppt. with 2 drops dil. HCl, and make a flame test. A brick-red flame test confirms calcium. Calcium flame test is not as definite as for Ba^{++}, Sr^{++}. Test the solution for Mg^{++} by adding 2 drops Na_2HPO_4, and allow to stand for 5 minutes. A white crystalline ppt. indicates Mg^{++}. If Mg^{++} is found here, it should also be found in Group V.

PRECIPITATE: $BaCrO_4$ (yellow). Wash the ppt. with H_2O until washings are colorless. Reject washings. Add 2 drops 6 *M* HCl. Dip a clean platinum wire in the solution, and then put the wire in a nonluminous flame. A green flame test confirms *barium.*

SOLUTION: Sr^{++}, CrO_4^{--}. Add 6 *M* NH_4OH dropwise until basic to litmus. Concentrate the solution to 0.5–1 ml. Add an equal volume of ethyl alcohol while stirring the mixture (5). Cool under the tap and let stand 5 minutes. A yellow ppt. indicates *strontium.* Cfg. Do not wash ppt.

PRECIPITATE: $SrCrO_4$. Add 10 drops $(NH_4)_2SO_4$ solution, and heat to boiling for about 1 minute in a water bath. Cool under the tap. Cfg. Wash ppt. of $SrSO_4$ with H_2O. Discard washings. Treat ppt. with 2 drops 6 *M* HCl, and make a flame test. A crimson color confirms *strontium* (6).

SOLUTION: Discard.

NOTES.

1. The concentration of OH^- is kept low by the presence of NH_4^+. Without this, the $[OH^-]$ would be so large that the product of $[Mg^{++}] \times [OH^-]^2$ would exceed the solubility product and thus ppt. $Mg(OH)_2$.

2.
$$MgCO_3 \quad K_{sp} = 2.6 \times 10^{-5}$$
$$BaCO_3 \quad K_{sp} = 4.9 \times 10^{-9}$$
$$CaCO_3 \quad K_{sp} = 1.7 \times 10^{-8}$$
$$SrCO_3 \quad K_{sp} = 4.6 \times 10^{-9}$$

If the concentration of Mg^{++} in the original sample is high, some $MgCO_3$ will probably ppt. here. It probably will not interfere with the tests for Ba^{++}, Sr^{++}, Ca^{++}, however. It should be tested for in this group as well as in Group V. The concentration of CO_3^{--} should be so regulated that all the $BaCO_3$ will be pptd. but $MgCO_3$ will not be—that is, $[Ba^{++}] \times [CO_3^{--}]$ should exceed its solubility product of 4.9×10^{-9}, but $[Mg^{++}] \times [CO_3^{--}]$ should not exceed its solubility product of 2.6×10^{-5}. Control of the hydroxyl ion concentration aids in this regard. $(NH_4)_2CO_3$ is so extensively hydroxyzed that a solution of it (in the absence of NH_4OH) does not contain enough carbonate ions to completely ppt. $BaCO_3$, $CaCO_3$, and $SrCO_3$.

$$CO_3^{--} + H_2O \leftrightharpoons HCO_3^- + OH^-$$

NH_4OH is added, increasing the OH^- concentration to shift this equilibrium to the left. Since too great a concentration of OH^- must also be prevented (see Note 1), the presence of NH_4^+ from some salt is essential.

3. Dry acetone or isopropyl alcohol may be used instead of dry (absolute) ethyl alcohol. These are solvents for $Ca(NO_3)_2$; $Ba(NO_3)_2$ and $Sr(NO_3)_2$ are relatively insoluble. H_2O in the solvent would dissolve all three nitrates, of course, and separation would not be effected.

Solubilities compared:

$Ca(NO_3)_2$ in dry C_2H_5OH	85.4 g/100 g
$Sr(NO_3)_2$ in dry C_2H_5OH	0.0062 g/100 g
$Ba(NO_3)_2$ in dry C_2H_5OH	0.0016 g/100 g
$Ca(NO_3)_2$ in dry isopropyl alcohol	2.6 g/100 g
$Sr(NO_3)_2$ in dry isopropyl alcohol	0.002 g/100 g
$Ba(NO_3)_2$ in dry isopropyl alcohol	0.0016 g/100 g

4.

$$BaCrO_4 \quad K_{sp} = 2.3 \times 10^{-10}$$
$$SrCrO_4 \quad K_{sp} = 3.8 \times 10^{-4}$$
$$CaCrO_4 \quad K_{sp} = 2.3 \times 10^{-2}$$

The method of separating Ba^{++} and Sr^{++} is based on the relative solubilities of the chromates. The CrO_4^{--} ion concentration is to be controlled so that $BaCrO_4$ is pptd., but strontium chromate is left in solution. The CrO_4^{--} ion concentration may be controlled by the H^+ ion concentration, as shown by the equation

$$2\ CrO_4^{--} + 2\ H^+ \rightleftharpoons Cr_2O_7^{--} + H_2O$$

If the concentration of H^+ is too small, the concentration of CrO_4^{--} will be high enough to precipitate $SrCrO_4$. If the concentration of H^+ increases, the concentration of CrO_4^{--} decreases, so that $BaCrO_4$ may not ppt. completely. The proper concentration of H^+ is maintained in an acetic acid–ammonium acetate buffer solution. Thus the acetate ions hold H^+ and CrO_4^{--} ions in reserve in the form of $HC_2H_3O_2$ and $Cr_2O_7^{--}$, respectively.

5. $SrCrO_4$, although more soluble than $BaCrO_4$, will ppt. if the H^+ is low (that is, chromate high), and if alcohol is added. (Practically all inorganic salts are less soluble in alcohol than in H_2O.) In this reaction more K_2CrO_4 reagent need not be added, since the excess is still present from the original addition of it.

6. If $BaCrO_4$ was not entirely separated from the strontium and calcium salts, the flame test at this point will show a flash of green before the more persistent crimson color of strontium.

D. Analysis of unknown solutions. After studying Group V, run a mixed unknown of Group IV and V. No separate unknown of this three-ion group is necessary.

Group V The Alkali Group

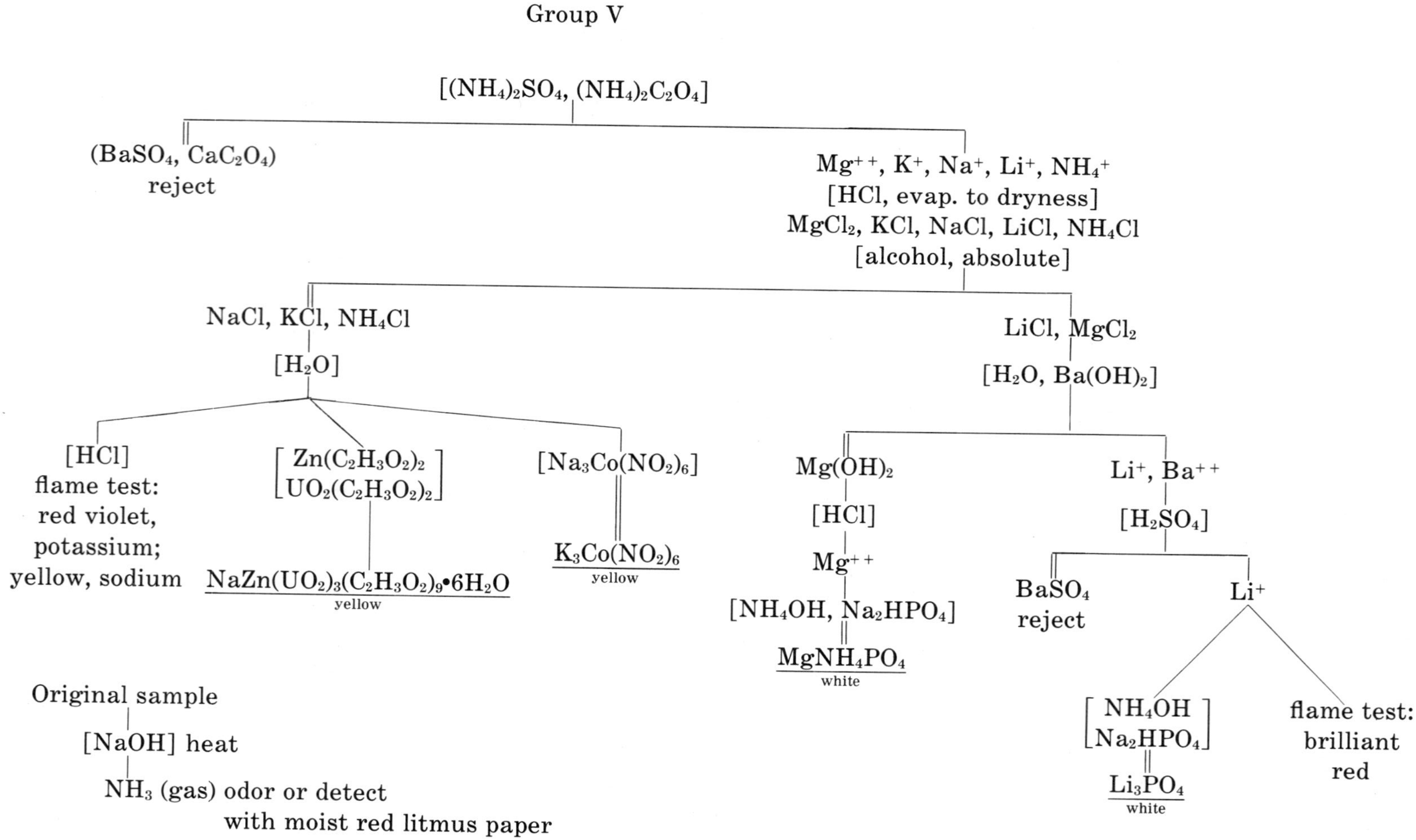

25
Group V
the alkali group

A. Chemistry of separations and tests.

1. The basis for separating Mg^{++} and Li^{+} from Na^{+} and K^{+} is the relatively greater solubility of $MgCl_2$ and LiCl in absolute alcohol. Below are the solubilities of the chlorides in grams per 100 g absolute ethyl alcohol at 20° C:

$MgCl_2$	LiCl	NaCl	KCl
5.6	24.3	0.065	0.034

   Isopropyl alcohol may be used in place of C_2H_5OH.

2. Mg^{++} may be separated from Li^{+} by precipitation of $Mg(OH)_2$ with $Ba(OH)_2$.

$$Mg^{++} + 2\ OH^{-} \rightarrow \underline{Mg(OH)_2}$$

   After solution of $Mg(OH)_2$ in acid, Mg^{++} is identified by precipitation of $MgNH_4PO_4$ in ammoniacal solution with Na_2HPO_4.

$$Mg^{++} + NH_4^{+} + PO_4^{3-} \rightarrow \underset{\text{white}}{\underline{MgNH_4PO_4}}$$

3. Li^{+} is identified by its brilliant red flame test, and also by precipitation of Li_3PO_4.

$$3\ Li^{+} + Na_2HPO_4 + OH^{-} \rightarrow \underset{\text{white}}{\underline{Li_3PO_4}} + 2\ Na^{+} + H_2O$$

4. Na^{+} and K^{+} may be identified in the presence of each other by precipitation reactions. K^{+} is identified by precipitation of yellow potassium cobaltinitrite, $K_3Co(NO_2)_6$, with a fresh solution of $Na_3Co(NO_2)_6$.

$$3\ K^{+} + Na_3Co(NO_2)_6 \rightarrow \underset{\text{yellow}}{\underline{K_3Co(NO_2)_6}} + 3\ Na^{+}$$

   Na^{+} may be precipitated with zinc uranyl acetate reagent as a finely divided yellow solid:

$$2\ Na^{+} + 6Zn(C_2H_3O_2)_2 \cdot UO_2(C_2H_3O_2)_2 + 12\ H_2O \rightarrow \underset{\text{yellow}}{\underline{2NaZn(UO_2)_3(C_2H_3O_2)_9 \cdot 6H_2O}} + 3Zn(C_2H_3O_2)_2 + Zn^{++}$$

   Both Na^{+} and K^{+} may be confirmed by flame tests. K^{+} gives a reddish violet flame when viewed through a cobalt glass filter (to filter out any sodium flame). Na^{+} gives a fluffy yellow flame which lasts several seconds if an appreciable

amount of sodium is present in the sample. A little practice enables one to differentiate between traces of sodium and appreciable amounts of the element.

5. NH_4^+ ion is detected by liberation of NH_3 gas by treatment with a strong base. For example:

$$NH_4^+ + OH^- \rightarrow NH_3\uparrow + H_2O$$

The ammonia gas may be detected by odor or by its effect on moist red litmus.

Equations for Group V

Magnesium

Sodium

Potassium

Ammonium

Lithium

B. Preliminary tests for Group V. None. You may proceed directly to the analytical procedures.

C. Analytical procedures for Group V. Make up a known solution using 4 drops each of test solutions of Mg^{++}, K^+, Li^+, Na^+, NH_4^+. Analyze according to Procedure 13.

PROCEDURE 13 *Analysis of Group V*

SOLUTION (from Procedure 12): Mg^{++}, Li^+,* K^+, Na^+, NH_4^+. Adjust the volume of solution to about 0.5 ml. If solution is not already basic, add dil. NH_4OH until basic. Add 1 drop $(NH_4)_2SO_4$ and 1 drop $(NH_4)_2C_2O_4$ (1). If ppt. forms, cfg. and reject ppt. Treat the solution according to directions below. If Li^+ might be present, treat according to Procedure A. If Li^+ is known to be absent, treat separate portions of the solution according to Procedure B.

A. If Li^+ is to be tested for: Place the solution in a small crucible or evaporating dish, add 2 drops 6 *M* HCl, and carefully evaporate to dryness. Treat residue with 2 ml absolute alcohol or isopropyl alcohol; break up the solid with a stirring rod and stir well to digest the residue. Transfer residue and solution to a small dry test tube, then cfg.

RESIDUE: NaCl, KCl, NH_4Cl. Dissolve in 2 ml H_2O. Use separate portions to test for Na^+ and K^+ according to Procedure B below.	**SOLUTION:** $MgCl_2$, LiCl. Add 1 ml H_2O, then 0.5 ml $Ba(OH)_2$. Heat to near boiling. Cfg.

PRECIPITATE: $Mg(OH)_2$. Dissolve with 3–4 drops dil. HCl. Add dil. NH_4OH until basic; then add 2 drops Na_2HPO_4. Allow to stand for a few minutes (2). Formation of a white crystalline ppt. ($MgNH_4PO_4$) proves *magnesium.*	**SOLUTION:** Li^+, Ba^{++}. Add 3 drops dil. H_2SO_4 to ppt. Ba^{++} as $BaSO_4$. Cfg. and discard ppt. Evaporate solution to about 1 ml. Use a few drops of solution for flame test. A brilliant red flame indicates Li^+. To the remainder of solution add NH_4OH until basic; add 2 drops Na_2HPO_4. A slow-forming white ppt. proves *lithium* (3).

B. If Li^+ is not present: Test for Mg^{++}, Na^+, and K^+ as follows:

TEST FOR Mg^{++}. To 6 drops of the solution in a test tube, add 2 drops dil. NH_4OH, followed by 2 drops Na_2HPO_4. Allow to stand for a few minutes (2). The formation of a white crystalline ppt. ($MgNH_4PO_4$) proves *magnesium.*

FLAME TESTS FOR Na^+ AND K^+. Evaporate several drops of the solution to dryness in a crucible or evaporating dish. Moisten residue with a drop of dil. HCl and make flame test. Observe the flame through a cobalt glass filter. A reddish violet flame through the glass indicates *potassium.* If the flame is a fluffy yellow (without filter) which lasts for several seconds, *sodium* is present (4). Confirm these tests by using the *original* sample. If appreciable amounts of sodium are present, a good flame test should be obtained on the *original* sample.

TEST FOR Na^+. To a few drops of solution in a test tube add 8 drops zinc uranyl acetate. Cool under the tap and allow to stand for 10 minutes. Cfg. A fine yellow ppt. confirms *sodium.*

* Lithium may be omitted at the option of the instructor.

TEST FOR K^+. Evaporate 6 drops of the solution to dryness in an evaporating dish. Add 2–3 drops dil. HNO_3 to the residue and again evaporate to dryness. Bake the residue for several minutes with a strong flame (5). Treat the residue with a drop of dil. $HC_2H_3O_2$ and 3–4 drops H_2O. Stir well and transfer to small test tube. Add a drop of freshly prepared sodium cobaltinitrite. Warm. A yellow ppt. forming slowly confirms *potassium.*

TEST FOR NH_4^+. The test for ammonium ion must be made on the *original* sample, since ammonium salts have been added in the analysis. To 3 drops of the original sample (or several crystals, if the sample is solid) in the bottom of a test tube, carefully add NaOH dropwise until basic (6). Place a moistened strip of red litmus over the edge of the tube so that about one-half of it is on the inside of the tube. Heat the tube cautiously. A blue color to the lismus proves *ammonium* ion. (Instead of using the litmus above, the solution may be heated to boiling and smelled. The odor of ammonia is usually very apparent if *ammonium* ion is present.)

NOTES.

1. Since $BaCO_3$ and $CaCO_3$ are slightly soluble in alkaline solution, small amounts of Ca^{++} and Ba^{++} may appear in the filtrate containing Group V. The purpose of adding $(NH_4)_2C_2O_4$ and $(NH_4)_2SO_4$ is to remove these ions as $BaSO_4$ and CaC_2O_4. If they were not removed, they would interfere with the test for Mg by pptn. as insoluble phosphates.

2. The ppt. of $MgNH_4PO_4$ forms slowly; consequently, the solution should be allowed to stand for 5–10 minutes.

3. Li_3PO_4 is slow to form and resembles small ice crystals. Be sure solution is basic.

4. Potassium and sodium should *always* be confirmed by flame tests on the *original* sample. Evaporate a few drops of solution to nearly dryness for the tests.

5. Ammonium salts are volatile and hence can be removed by baking the residue of salts. If not removed, the ammonium salts interfere with the test for K^+. Freshly prepared $Na_3Co(NO_2)_6$ is imperative. Oxidation and decomposition of sodium cobaltinitrite occurs as the reagent stands. It is best to test reagent for effectiveness with a known K^+ solution before using with unknown.

6. Carefully avoid touching the test tube with the pipet used to insert NaOH. If NaOH comes in contact with the litmus, a false test for NH_4^+ will be obtained.

D. Analysis of unknown solution over Groups IV–V. Obtain from the instructor an unknown containing ions of Groups IV–V only. Use 0.5 ml for analysis. Analyze according to Procedures 12 and 13. Record results in a diagrammatic scheme on the opposite page.

E. Analysis of general unknown. Obtain from your instructor an unknown over Groups I–V. Analyze according to procedures for cation analysis. Use 0.5 ml for analysis. Develop a diagrammatic scheme on next page. In the space below, write equations for reactions of all ions found. Report the results to your instructor.

NOTE TO INSTRUCTOR: More than one general unknown may be assigned if time permits.

Part five

Analysis of alloys, salts, and commercial substances

26
Analysis of alloys

The product obtained by melting together two or more metals and allowing the mixture to solidify is called an *alloy*. In the solid state, just as in liquids, elements vary in solubility in each other, from very soluble to very insoluble. Thus we may have in the solid form

1. A solid solution of one or more metals, one in another, with any excess above the amount required for saturation present as crystals imbedded in the solid solution.
2. Elements almost insoluble in each other in solid form, resulting in a close-packed mixture of distorted crystals.
3. The formation of compounds between the metals on cooling.

Common alloys.

TRADE NAME	COMPOSITION, %
Aluminum bronze	90 Cu, 10 Al
Babbitt	90 Sn, 7 Sb, 3 Cu
Bearing bronze	82 Cu, 16 Sn, 2 Zn
German silver	60 Cu, 25 Zn, 15 Ni
Gold coinage	90 Au, 10 Cu
Magnalium	90 Al, 10 Mg
Manganese bronze	95 Cu, 5 Mn
Nickel coinage	75 Cu, 25 Ni
Pewter	85 Sn, 6 Bi, 7 Cu, 2 Sb
Red brass	90 Cu, 10 Zn
Silver coinage	90 Ag, 10 Cu
Solder	50 Pb, 50 Sn
Type metal	82 Pb, 15 Sb, 3 Sn
Yellow brass	67 Cu, 33 Zn
Wood's metal	50 Bi, 25 Pb, 12.5 Sn, 12.5 Cd

It is evident that the composition of alloys may vary within wide limits; certain metals may be present in very small percentages in certain alloys, in others, a high percentage of these metals may be present. In the foregoing analysis of known solutions of cations, a sample was taken which contained 0.5 to 1 mg of each ion to be tested. Such an ideal case of having each of the ions present in about equal quantities is rarely found in the case of alloys. Thus one is confronted with problems of choosing the proper size of sample so that all elements will be found, and yet not so much of any

one element to hinder "clean" or "complete" separation from other elements. For example, if a certain bronze alloy containing 82% Cu, 16% Sn, and 2% Zn is to be analyzed, the sample should be large enough so that zinc will be found. However, since the copper content is 41 times as great as that of zinc, too large a sample must be avoided. If a 20 mg sample of this alloy is taken at the start of the analysis, the quantity of zinc to be precipitated in Group III (0.4 mg) will be comparable with that in 1 drop of Zn^{++} t.s. (*In general, 20 mg is a suitable size sample for an unknown alloy.*) In this same sample 16.4 mg of Cu will be precipitated in Group II (as compared with 0.5 mg in a known.)

Thus the results of the analysis may, and should rightly, be reported on a semi-quantitative basis. An attempt should be made to classify an element as a *major ingredient*, a *minor ingredient* (a small percentage), or a *trace*. Impurities in reagents sometimes give trace tests.

Nitric acid is probably the most effective reagent in attacking alloys, since it dissolves all metals except gold and platinum to form soluble nitrates. Tin and antimony, however, form insoluble oxides:

$$Cu + 4\ HNO_3 \rightarrow Cu(NO_3)_2 + 2\ NO_2 + 2\ H_2O$$

$$Fe + 6\ HNO_3 \rightarrow Fe\ (NO_3)_3 + 3\ NO_2 + 3\ H_2O$$

$$Sn + 4\ HNO_3 \rightarrow \underline{SnO_2} + 4\ NO_2 + 2\ H_2O$$

$$2\ Sb + 10\ HNO_3 \rightarrow \underline{Sb_2O_5} + 10\ NO_2 + 5\ H_2O$$

Ferrous and non-ferrous alloys. The alloys listed above may be classed as non-ferrous alloys. Ferrous alloys are for the most part specialty steels, and the elements commonly alloyed with Fe are Cr, Mn, W, Mo, V, Ni, Co, Si, and C. The instructor may designate whether an unknown alloy is a ferrous or non-ferrous one. Ferrous alloys are hard to dissolve in HNO_3 and should be dissolved in HCl. Consequently, if little action is noted with HNO_3, the alloy should be washed free of HNO_3 and treated with conc. HCl. In the latter case, tin and antimony will appear in solution with other cations, and will be tested for in the regular procedures for cation analysis.

Analysis of an unknown alloy. Obtain an unknown alloy from your instructor and analyze according to Procedure 14. He will probably tell you whether it is a ferrous or non-ferrous alloy. Record results on opposite page.

PROCEDURE 14 *Analysis of Alloy*

Place about 20 mg of the alloy in the form of turnings or filings in a small beaker. Follow Procedure A for a non-ferrous alloy and Procedure B for a ferrous alloy.

A. *Non-ferrous.* Add 3 ml 6 *M* HNO_3 (1),(2), cover with a watch glass, and heat in hood until brown fumes are no longer evolved. (Use more HNO_3 as necessary.)
(a) If solution is complete: Place solution in an evaporating dish and boil off excess HNO_3 (use hood). The material can be taken to almost dryness – to crystal formation. Avoid further heating as nitrates decompose to water-insoluble oxides. Add 5 ml cold H_2O and 1 ml conc. HCl. If ppt. forms, separate ppt. and analyze the ppt. for the silver group, Procedure 4. Evaporate the filtrate (or the solution in case no ppt. was formed with HCl) almost to dryness to remove excess acid. Take up residue in 5 ml H_2O and analyze as usual for the cations of Groups II and III.
(b) If solution is incomplete: A white residue remains. (A white residue here may be hard to see against the evaporating dish background.) Add 5 ml H_2O, stir, and centrifuge. Decant the solution and later analyze it for cations according to Procedure 4. Wash the white residue of possible SnO_2 and Sb_2O_5 at least three times with dil. HNO_3. Discard washings. Place residue in an evaporating dish; add 2 ml conc. HCl and heat to boiling to dissolve antimony and tin oxides. Analyze the solution possibly containing $SnCl_4$ and $SbCl_3$ as follows: Add 0.2 g oxalic acid crystals and 5 drops H_2O to 5 drops of the solution. Add H_2S (thioacetamide) to the hot solution. Reddish orange Sb_2S_3 indicates antimony. Separate the Sb_2S_3, if any, from the solution. Boil the solution until H_2S is removed. Add dil. HCl if necessary to keep volume constant. Add a pinch of test lead or a piece of Mg ribbon to the solution. Boil 2–3 minutes. Cfg. and discard any residue. Cool the mixture and pour it into a solution of $HgCl_2$. White to grey ppt. indicates tin.

B. *Ferrous.* Add 8–10 drops 6 *M* HCl and warm until action ceases or until solution is complete. Cfg. and reject any small residue (2). Use the solution for analysis of Cation Groups II and III (3).

NOTES.

1. Unless the alloy decomposes H_2O, no alkali or alkaline earth metals are present. Calcium, barium, strontium, sodium, and potassium are seldom included in alloys.

2. C, P, As, and Si are sometimes found in alloys. As and P will dissolve in HNO_3, forming arsenic and phosphoric acids. C and Si will be an insoluble residue, and should not be mistaken for Sb and Sn. Do not use too much HNO_3 or evaporation will require too much time.

3. Group I metals cannot be present since Hg_2Cl_2, $PbCl_2$, and AgCl are insoluble. The excess HCl should be removed by boiling or by neutralization with NH_4OH such that *p*H may be adjusted for Group II precipitation with sulfide ion. One does not usually find Group II metals (Cu, Bi, Cd, Sb, Hg, Sn, As) in a ferrous alloy, consequently the analytical procedures for ferrous alloys will normally start with Group III.

27
Analysis of salts and mixtures

The analysis of a salt or mixture of salts consists of the following steps:

1. Physical examination of sample
2. Solution of sample
3. Cation analysis
4. Anion analysis

Physical examination. If an unknown is a solid, a careful observation of certain physical properties may yield valuable information about its composition. Does the substance appear homogeneous and uniform in composition? Is it crystalline or noncrystalline? Is it metallic or salt-like? Does the color indicate certain elements? These questions may be answered by careful inspection of the sample. Certain cations and anions have characteristic colors which may be indicative of the composition. Below are listed some of the more common colored ions:

Cu^{++}	blue	CrO_4^{--}	yellow
Ni^{++}	green	$Cr_2O_7^{--}$	orange-red
Co^{++}	pink to blue	MnO_4^-	purple
Fe^{3+}	yellow (with Cl^-)	Mn^{++}	pale pink
Cr^{3+}	green	Fe^{++}	pale green

Flame test OH^- test

Selection of a solvent. Before analyzing a solid unknown for cations and anions, it is necessary to dissolve the sample. A pinch of the sample no larger than the head of a pin should be treated with 2–3 ml of various solvents to determine which solvent to use in the preparation of a solution for cation analysis. The order of solvents is: (1) cold then hot water; (2) dil. HCl (hot if necessary); (3) dil. HNO_3 (hot if necessary); (4) hot conc. HCl; (5) hot conc. HNO_3; (6) hot conc. HCl and hot conc. HNO_3. Test the solubility of the sample in these solvents in the order listed until a suitable solvent is found. In determining the solvent to use, allow plenty of time for solution; if the solvent appears to act slowly on the sample, allow more time for solution. Certain mixtures may be partially soluble in one solvent or another; it is best to find a solvent which will completely dissolve the sample, and thus obtain only one solution for analysis. Certain substances, such as $PbSO_4$, AgCl, $BaSO_4$, are insoluble in the solvents listed above and require fusion with Na_2CO_3 for solution. Substances requiring special treatment for solution will not be included in this course.

After selecting a suitable solvent, dissolve a portion about the size of a small pea in as little solvent as possible and analyze for cations according to Procedure 4.

Analysis for cations. Ordinarily the analysis for c
anions, since, on the basis of the cation analysis, eliminatior
indicated. For example, if the anions CrO_4^{--} or AsO_4^{3-} are pr
elements will be found in the cation analysis. Obviously, if
found in the cation analysis, the anions could not be present,
need not be made. Also, if the sample is a solid, certain ani
the basis of solubility considerations and the cation anal
which dissolves in water is found to contain Ba^{++}. Since the
SO_4^{--}, CO_3^{--}, SO_3^{--}, BO_3^{3-}, AsO_4^{3-}, CrO_4^{--}, and PO_4^{3-} are insoluble in water,
anions may be eliminated as possibilities and anion tests need by made only for the anions which yield soluble barium salts—that is, Cl^-, Br^-, I^-, S^{--}, NO_3^-, NO_2^-, $C_2H_3O_2^-$.

Analysis for anions. If cations other than the alkalies have been found in the cation analysis, they may interfere with the group tests and specific tests for the anions. Hence it will be necessary to prepare a solution free of interfering cations before making group tests and specific tests for anions. The preparation of a solution for anion analysis may be accomplished by Procedure 15. Use this prepared solution for making group tests and specific tests according to Procedures 2 and 3.

PROCEDURE 15 *Sample for Anion Analysis*

PREPARATION OF A SOLUTION FOR ANION ANALYSIS. Boil a small portion (0.2–0.3 g) of the solid unknown with 5 ml conc. Na_2CO_3 solution. Filter, and discard the ppt. Acidify the filtrate with HNO_3 and boil until evolution of CO_2 has ceased. Use this solution in group tests and specific tests for anions (except for carbonate, nitrate, nitrite, sulfide, and sulfite) (1).

$$\text{Cation–Anion} + Na_2CO_3 \rightarrow \underline{\text{Cation–Carbonate}} + \text{Na–Anion}$$

$$\text{Na–Anion} + HNO_3 \rightarrow NaNO_3 + H_2O + CO_2\uparrow$$
$$+ H_2S\uparrow \text{ or } S$$
$$+ SO_2\uparrow$$
$$+ \text{H–Anion solution}$$

If interfering cations are absent from the unknown, a solution for anion analysis may be prepared simply by dissolving a portion of the solid in water.

NOTES.

1. HNO_3 would destroy CO_3^{--}, NO_2^-, SO_3^{--}, and of course a solution to which HNO_3 has been added should not be tested for NO_3^-. The test for these ions is on the *original* sample.

Analysis of unknowns. Each student should analyze at least two single-salt samples and one mixed-salts sample. Obtain samples from the instructor and record all information on the record sheets provided.

Salt Analysis.

Number of sample ____________

ysical properties of sample

Color? ____________ Crystalline or noncrystalline? ____________

Homogeneous? ____________ Soluble in H_2O? ____________

If insoluble in H_2O, in what solvent does it dissolve? ____________

Cation analysis

Group I ____________ Group II ____________ Group III ____________

Group IV ____________ Group V ____________ NH_4^+ ____________

Anion analysis

H_2SO_4 treatment (what happened?) ____________

On the basis of H_2SO_4 treatment and solubility considerations, can you eliminate any anions as possibilities? If so, list them:

____

Group tests for anions

Sulfate group ____________ Halide group ____________

Specific tests for anions: If you did not make certain specific tests, indicate why.

Sulfate ____________ Chloride ____________

Sulfite ____________ Bromide ____________

Carbonate ____________ Iodide ____________

Chromate ____________ Sulfide ____________

Borate ____________ Acetate ____________

Arsenate ____________ Nitrate ____________

Phosphate ____________ Nitrite ____________

Report on unknown:

Record of Salt Analysis.

Number of sample............................

Physical properties of sample

Color?............................ Crystalline or noncrystalline?............................

Homogeneous?............................ Soluble in H_2O?............................

If insoluble in H_2O, in what solvent does it dissolve?............................

Cation analysis

Group I............................ Group II............................ Group III............................

Group IV............................ Group V............................ NH_4^+............................

Anion analysis

H_2SO_4 treatment (what happened?)............................

On the basis of H_2SO_4 treatment and solubility considerations, can you eliminate any anions as possibilities? If so, list them:

....

Group tests for anions

Sulfate group............................ Halide group............................

Specific tests for anions: If you did not make certain specific tests, indicate why.

Sulfate............................ Chloride............................

Sulfite............................ Bromide............................

Carbonate............................ Iodide............................

Chromate............................ Sulfide............................

Borate............................ Acetate............................

Arsenate............................ Nitrate............................

Phosphate............................ Nitrite............................

Report on unknown:

Record of Salt Analysis.

Number of sample..............................

Physical properties of sample

Color?.. Crystalline or noncrystalline?..............................

Homogeneous?.................................... Soluble in H_2O?..

If insoluble in H_2O, in what solvent does it dissolve?..

Cation analysis

Group I.................................... Group II.................................... Group III..............................

Group IV.................................. Group V.................................... NH_4^+..

Anion analysis

H_2SO_4 treatment (what happened?)..

On the basis of H_2SO_4 treatment and solubility considerations, can you eliminate any anions as possibilities? If so, list them:

....

Group tests for anions

Sulfate group.. Halide group..

Specific tests for anions: If you did not make certain specific tests, indicate why.

Sulfate.. Chloride..

Sulfite.. Bromide..

Carbonate.. Iodide..

Chromate.. Sulfide..

Borate.. Acetate..

Arsenate.. Nitrate..

Phosphate.. Nitrite..

Report on unknown:

28
Analysis of commercial substances

Procedures have been outlined for identification of the *common metals* (singly or in admixture) under alloy analysis coupled with cation analysis.

Many *pharmaceuticals* and *industrial substances* are salts or salt mixtures, and may be tested for as such. Anion analysis identifies acids. Qualitative cation testing of solutions made from metal oxides or hydroxides identifies these types of substances.

Separation of organic and inorganic matter. The scope of this text is inorganic analysis. Analysis of organic compounds is a separate branch of chemistry. At times, however, *a mixture of inorganic and organic matter* is encountered. The presence of organic matter in a sample may usually be detected by heating a small portion of the solid sample in a small test tube. Charring and burnt odor accompanies decomposition of most organic matter. By heating a larger portion of organic-inorganic mixture to a high temperature in a crucible, carbonaceous material may be volatilized to leave an inorganic residue. Consideration must be given to the fact that some inorganic matter can be vaporized, partly decomposed, or sublimed [for example, As_2O_3, $Cu(NO_3)_2$, $CaCO_3$, Hg_2Cl_2].

Detection of common inorganic substances in foods may be carried out after removal of organic matter. In testing food for possible inorganic materials that might volatilize or decompose in the ignition process, a wet oxidation of the organic matter from inorganic may be carried out. Put a 1 g sample of the dried and mascerated mixture containing organic matter in a casserole or evaporating dish, and give vigorous digestion with a hot $HNO_3-H_2SO_4$ mix. Finally, heat to eliminate HNO_3 to the point of obtaining SO_3 fumes. Cool the resulting solution and pour it into cold water to give a solution for inorganic analysis.

In separating some organic from inorganic matter, use is made of solvents for organic matter, such as CCl_4, ether, turpentine, and gasoline. Subsequent to such organic removal, inorganic material may be separately analyzed.

Analyzing minerals. Rocks and minerals are most commonly identified (as studied in petrography and mineralogy) from physical properties of the minerals. The elements of qualitative chemical analysis are often employed in definitely identifying minerals not readily discerned from physical properties. Since most minerals are water insoluble, the main problem in qualitative testing is dissolving them. Most minerals are oxides, mixed oxides, or salts, and acids will dissolve many of them.

The most common acidic radicals in rocks are siliceous. Ground silicate minerals are first fused with excess Na_2CO_3. The fusion residue may be digested with hot water and then filtered. The residue may be dissolved in acid for cation analysis.

$CaO \cdot FeO \cdot Al_2O_3 \cdot XSiO_2$ (a silicate mineral)
fuse with Na_2CO_3
digest with water

$CaCO_3$, $FeCO_3$, $Al(OH)_3$ | $Na_2O \cdot SiO_2$, Na_2CO_3

[HCl]
solution for cation analysis

Siliceous rocks may be decomposed for analysis by treating the ground rock in a platinum crucible with HF. SiF_4 volatilizes and the fluoride metal salt residue can be dissolved for cation analysis.

Part six

Problems, exercises, and appendix

29
Principles and methods of instrumental analysis

I. Emission spectrography. Bunsen and Kirchhoff, many years ago, discovered that the spectra of flames colored by certain metallic salts are characteristic of the metals. Visually one can detect the red of strontium, green of barium or copper, yellow of sodium in a flame. Atoms or ions that have been excited emit their excess energy in the form of radiant energy of definitive wavelengths. The wavelengths theoretically relate to the diverse electron drops in energy levels from higher energy levels to which they had been excited. A high voltage spark, alternating or direct current arc, can excite atomic species in an emission spectrometer. A complex of prisms or diffraction gratings orient the wavelengths which can be recorded on a ribbon-like photographic film. Patterns of the line spectra of a known element can be compared to the spectra of an unknown to discern its presence or absence. Comparative intensity of the spectra of different elements present gives a measure of quantity.

Flame photometry eliminates the use of arc or spark sources of exciting energy; the sample to be analyzed in solution form is sprayed into a flame. The light of the flame enters a monochromator to select the desired wave length region. Intensity of the isolated spectral band for a given element may be measured by use of a photocell or some type of amplifying electronic meter.

Infrared spectrophometry aims at identification of bonding arrangements (for example, C–H, C–N, S–H, etc.). Different modes of vibration and rotation of units are caused by excitation energy. Infrared absorption spectra can be used to identify the components of any substance whether it be pure or impure. It is more applicable to organic chemistry and use of organic solvents for unknowns. This is because water, the best solvent for inorganic compounds, gives a strong absorption spectra beyond 1.5 μ, and inorganic compounds have broad absorption bands in contrast to narrow bands for the organic.

X-ray methods. Absorption of energy may shift an electron to an outer or higher energy level. Fall of the electron back to inner levels is accompanied by X rays—of a specific series of wavelengths according to the nature of the atom. The wavelength from an unknown source element is determined by using a crystal with a known spacing as a diffraction grating. Tables are available for the wavelengths of X rays from elements. This instrumentation method of qualitative analysis is especially good for the detection of rare elements or for the identification of post-uranium elements.

Mass spectrometer. In a mass spectrograph, ions are brought to a focus according to the ratio of mass charge, m/e. Exit slits are used to segregate separated isotopes, which are qualitatively or quantitatively determined. Photographic and electronic analyzers utilize the respective m/e beams of such specific ions as, for example, $^{40}Ca^{++}$ of $40m/2e$ and $^{56}Fe^{3+}$ of $56m/3e$.

II. Chromatography. *Gas chromatography* uses the principle of differential adsorption of gaseous or volatile substances as they pass through a tube-like column containing some solid or liquid adsorbent on the surface of a granular inert solid. Diverse substances are separated as they are selectively adsorbed in different regions of the adsorbent column. The separated substance may then be more easily identified and quantified than when it was part of a mixture.

Solution chromatography and adsorption columns. Selective rate of adsorption furnishes a basis for very many otherwise difficult separations prior to analysis of separated units from a mixture. Material for adsorption may be as simple as filter paper or as complex as an anion- or cation-adsorbing resin in a small or large sized column. Examples are: (1) a mixture of dye materials in a solution may be arranged to rise by capillary action into an end-immersed piece of filter paper and the colors of the dyes will collect in bands according to their traverse rates, and (2) a strong $HCl-H_2O$ solution of $NiCl_4^{--}$ and $CoCl_4^{--}$ will give adsorption of both ions on an anionic granular resin column. A gradual dilution of the acid through the tower will selectively release Ni^{++} to separate it from Co^{++}.

Separations of the very similar rare earth elements for identification (and also for industrial selective uses) are made by adsorption-column techniques.

III. Nuclear Magnetic Resonance. Most analytical methods relate to chemical or physical electron behavior. About half of the known atomic nuclei behave as if they are spinning entities in a fixed magnetic field subject to alteration in motion by an alternating magnetic field. Nuclear magnetic resonance spectroscopy aims at identifying and quantifying nuclear species by instrumentation. Much as a gyroscope resists tilt under gravity pull, a spinning nucleus resists tilt, but a strong external magnetic field causes a slow rotational spin in the direction of the external field in addition to its own spin about axis.

The total spin aspects of nuclei differ as the nuclei fall into three classes: (1) those with even, equal numbers of neutrons and protons, ${}^{12}_{6}C$, ${}^{16}_{8}O$, of zero spin; (2) those with either an odd number of protons or neutrons; (3) those in which both the number of protons and neutrons are odd. The strength of the nuclear magnetic moment is different between isotopes, a specific value for each allows for separation and identification. The axis of rotation of a spinning nucleus generating a small magnetic field is said to be precessed (caused to wobble in rhythm) by an imposed alternating magnetic field.

IV. Miscellaneous. Such additional phenomena as radioactivity, fluorescence, thermal and electrical conductivity, electrode potential, and biochemical effects may also be utilized for instrumental analysis.

30 Problems and exercises

Problems and exercises are arranged to conform to the usual topical sequence of a qualitative analysis course. Answers are supplied, but the student should refer to the appropriate section of the manual for the basic information needed to complete the exercises.

Composition of solutions.

1. Determine the number of moles in 100 g of the following:

   (a) HCl (d) $Fe_2(SO_4)_3$
   (b) H_3AsO_4 (e) $NH_4C_2H_3O_2$
   (c) $ZnCl_2$ (f) $Cr(OH)_3$

   Ans. (a) 2.74 (d) 0.25
   (b) 0.70 (e) 1.3
   (c) 0.73 (f) 0.97

2. Determine the number of equivalents in 100 g for each of the compounds in the previous exercise.

   Ans. (a) 2.74 (d) 1.5
   (b) 2.1 (e) 1.3
   (c) 1.46 (f) 2.9

3. Calculate the gram atoms of the metallic element contained in the weights of the following compounds:

   (a) 31.2 g Ag_2SO_4 (b) 12.27 g $BaCl_2{\cdot}2H_2O$

   Ans. (a) 0.2
   (b) 0.05

4. How many milligrams of NaOH are required to react completely with 5.6 mg of:

   (a) Fe^{++} (b) Fe^{3+}

   Ans. (a) 8.0 (b) 12.0

5. How many grams are contained in:

   (a) 1 g ion of Hg^{++} (b) 0.05 g ion of Cu^{++} (c) 2 g ions of PO_4^{3-}

   Ans. (a) 200.6 g
   (b) 3.177 g
   (c) 190 g

6. Indicate the weights needed for 0.1 g ion quantities of each of the following:

(a) Cd^{++} (b) SO_4^{--} (c) NO_3^- (d) Hg_2^{++}

Ans. (a) 11.2 g (c) 6.2 g
(b) 9.6 g (d) 40.1 g

7. 49.5 g of $SrBr_2$ is dissolved in 1 liter of H_2O (Sr = 87.5, Br = 80).

(a) 49.5 g equals how many gram moles of $SrBr_2$?
(b) If 100% ionized, it would form what gram ion content of Sr^{++}, and what gram ion content of Br^-?
(c) If the $SrBr_2$ is 60% ionized, what is the gram mole concentration? What is the gram ion concentration of Sr^{++} and Br^-?

Ans. (a) 0.2 g mole $SrBr_2$
(b) 0.2 g ion Sr^{++}
0.4 g ion Br^-
(c) 0.08 g mole $SrBr_2$
0.12 g ion Sr^{++}
0.24 g ion Br^- per liter of solution

8. $Al_2(SO_4)_3$ dissolves in H_2O at 20°C to the extent of 17.1 g/liter of solution.

(a) What is the gram mole solubility per liter?
(b) What is the gram ion Al^{3+} solubility per liter?
(c) What is the gram ion SO_4^{--} solubility per liter?

Ans. (a) 0.05 (c) 0.15
(b) 0.10

9. 91.5 g of CdI_2 is dissolved to make 1 liter of solution.

(a) 91.5 g equals how many gram moles of CdI_2 (Cd = 112.4, I = 126.9)?
(b) If 100% ionized, it would form what gram ion content of Cd^{++} and I^-?
(c) If the CdI_2 is only 60% ionized, what is the gram ion and gram mole concentration in the solution?

Ans. (a) 0.25 g mole CdI_2
(b) 0.25 g ion Cd^{++}
0.50 g ion I^-
(c) 0.10 g mole CdI_2
0.15 g ion Cd^{++}
0.30 g ion I^- per liter of solution

10. Calculate the molarities and normalities of the following solutions:

(a) 0.146 g HCl in 100 ml of solution
(b) 2.45 g H_2SO_4 in 200 ml of solution

(c) 49 g H_2SO_4 in 4 liters of solution
(d) 2 g NaOH in 5 liters of solution
(e) 3.7 g $Ca(OH)_2$ in 400 ml of solution
(f) 32.5 g $FeCl_3$ in 500 ml of solution
(g) 427 g $Ga_2(SO_4)_3$ in 10 liters of solution

Ans. (a) 0.04 *M*; 0.04 *N*
(b) 0.125 *M*; 0.25 *N*
(c) 0.125 *M*; 0.25 *N*
(d) 0.01 *M*; 0.01 *N*
(e) 0.125 *M*; 0.25 *N*
(f) 0.4 *M*; 1.2 *N*
(g) 0.1 *M*; 0.6 *N*

11. Calculate the weight of solute (in grams) necessary for each of the following solutions:

(a) 250 ml of 0.1 *M* LiOH
(b) 2 liters of 0.05 *M* H_2SeO_4
(c) 2 liters of 0.05 *N* H_2SeO_4
(d) 100 ml of 0.1 *N* $Cr_2(SO_4)_3$
(e) 50 liters of 3 *N* $Cr_2(SO_4)_3$
(f) 5 liters of 1 *M* H_3AsO_4

Ans. (a) 0.6 g (d) 0.653 g
(b) 14.5 g (e) 9800 g
(c) 7.25 g (f) 710 g

12. Calculate the molarity of each of the following solutions of slightly soluble substances:

(a) AgCN, 2.2×10^{-4} g/liter
(b) As_2S_3, 5.2×10^{-4} g/liter
(c) $BaSO_4$, 2.4×10^{-3} g/liter
(d) PbI_2, 0.68 g/liter

Ans. (a) 1.6×10^{-6}
(b) 2.1×10^{-6}
(c) 1.0×10^{-5}
(d) 1.5×10^{-3}

13. Using specific gravities and percentage composition given below, calculate the approximate normality of each of the following concentrated laboratory reagents:

	Specific gravity	*Percent by weight*
(a) HCl	1.2	38
(b) HNO_3	1.42	70
(c) H_2SO_4	1.84	96
(d) NH_4OH	0.9	60

Ans. (a) 12.5 (c) 36
(b) 16 (d) 15

14. Calculate the volume of 0.1 *N* HCl necessary:

(a) to exactly neutralize 50 ml of 0.04 *N* KOH.
(b) to react with 2.12 g Na_2CO_3.
(c) to neutralize 100 ml 0.01 *N* $Ca(OH)_2$.
(d) to react with 5 liters of 0.4 *N* $AgNO_3$.

Ans. (a) 20 ml (c) 10 ml
(b) 400 ml (d) 20 liters

15. 100 ml of 0.05 *N* H_2SO_4 was neutralized with 0.1 *N* KOH, and the resulting solution was evaporated to dryness.

(a) What volume of 0.1 *N* KOH was required?
(b) What weight of K_2SO_4 was obtained on evaporation?

Ans. (a) 50 ml (b) 0.435 g

16. Calculate the volume of 0.2 *N* H_2SO_4 which can be prepared from 25 ml of 36 *N* H_2SO_4.

Ans. 4.5 liters

Exercises with exponentials.

1. Express the following in conventional exponential form:

(a) 10,000 (g) 52.01
(b) 400 (h) 45000
(c) 22400 (i) 0.069
(d) 0.005 (j) 96500
(e) 0.000032 (k) 602,000,000,000,000,000,000,000
(f) 625.2

Ans. (a) 10^4 (g) 5.201×10^1
(b) 4×10^2 (h) 4.5×10^4
(c) 2.24×10^4 (i) 6.9×10^{-2}
(d) 5×10^{-3} (j) 9.65×10^4
(e) 3.2×10^{-5} (k) 6.02×10^{23}
(f) 6.252×10^2

2. Express the following exponentials in decimal form:

(a) 5.0×10^4 (d) 4.17×10^{-2}
(b) 3.0×10^{-3} (e) 9.80×10^{-5}
(c) 6.8×10^6 (f) 4.62×10^0

Ans. (a) 50000 (d) 0.0417
(b) 0.003 (e) 0.000098
(c) 6800000 (f) 4.62

3. Perform the indicated operations.

(a) $5 \times (3 \times 10^3)$

(b) $(1.3 \times 10^2) \times (6.0 \times 10^4)$

(c) $(6.4 \times 10^{-5})(2.5 \times 10^4)$

(d) $(8.2 \times 10^{-3})(4.0 \times 10^{-4})$

(e) $\dfrac{6.8 \times 10^6}{3.4 \times 10^6}$

(f) $\dfrac{3.6 \times 10^4}{1.2 \times 10^{-2}}$

(g) $\dfrac{3.2 \times 10^{-6}}{8.0 \times 10^3}$

(h) $\dfrac{(8.0 \times 10^{-4})(6.5 \times 10^{-2})}{1.3 \times 10^4}$

(i) $\dfrac{4.8 \times 10^{-6}}{(1.2 \times 10^4)(4 \times 10^{-10})}$

(j) $\dfrac{(3.6 \times 10^{-4})(1.5 \times 10^3)}{(9.0 \times 10^8)(3.0 \times 10^{-7})}$

(k) $\sqrt[4]{160000}$

(l) $\sqrt{1.69 \times 10^6}$

(m) $\sqrt{14.4 \times 10^{-11}}$

(n) $(2 \times 10^3)^3$

(o) $(5 \times 10^{-4})^2$

(p) $\dfrac{(2 \times 10^3)^4}{6.4 \times 10^2}$

(q) $\dfrac{(8 \times 10^2)^3(4 \times 10^{-4})^2}{(4 \times 10^1)^4(2 \times 10^{-4})^3}$

Ans.

(a) 15,000	(j) 2×10^{-3}
(b) 7.8×10^6	(k) 20
(c) 1.6	(l) 1.3×10^3
(d) 3.28×10^{-8}	(m) 1.2×10^{-5}
(e) 2.0	(n) 8×10^9
(f) 3×10^6	(o) 2.5×10^{-7}
(g) 4×10^{-10}	(p) 2.5×10^{10}
(h) 4×10^{-9}	(q) 2×10^9
(i) 1.0	

Ionization constants of acids and bases.

1. From the following ionization constants (K_a), indicate the strongest and weakest acids.

(a) Boric acid	7.3×10^{-10}	(e) Thiocetic acid	4.7×10^{-4}
(b) Hypobromous acid	2.0×10^{-9}	(f) Valeric acid	1.5×10^{-5}
(c) Periodic acid	2.3×10^{-2}	(g) Glutaric acid	3.4×10^{-4}
(d) Chromic acid	1.8×10^{-1}	(h) Cyanoacetic acid	3.7×10^{-3}

Ans. (d) and (a)

2. Calculate the H^+ ion concentration of 0.1 *M* formic acid ($HCHO_2$) solution (K_a for formic acid $= 1.8 \times 10^{-4}$).

Ans. 4.2×10^{-3}

3. Calculate the value of K_a for 0.1 M HNO_2 which is 7% ionized.

Ans. 5×10^{-4}

4. A 0.5 M solution of the base MOH, is found to be 1×10^{-4}% ionized. Calculate K_b.

Ans. 5×10^{-13}

5. A 1.0 M solution of a certain base, MOH, is 15% ionized into M^+ and OH^-. Calculate K_b.

Ans. 2.6×10^{-2}

6. The base ionization constant for hydrazine (N_2H_4) is 2.5×10^{-5}. What is the OH^- concentration in 1.0 M solution?

$$N_2H_4 + H_2O \rightleftharpoons N_2H_5^+ + OH^-$$

Ans. 5×10^{-3}

7. A weak acid, HA, is 2.0% ionized in 0.2 M aqueous solution. Calculate K_a.

Ans. 8.2×10^{-5}

8. K_a for $HPO_3 = 5.0 \times 10^{-5}$. Calculate the concentration of H^+ in 0.005 M solution.

Ans. 5×10^{-4}

9. Calculate the OH^- ion concentration in 0.1 M NH_4OH solution (K_b for $NH_4OH = 1.8 \times 10^{-5}$).

Ans. 1.3×10^{-3}

10. Underscore the most abundant species (omit H_2O) and encircle the least abundant species in an aqueous solution made by dissolving 1 mole of $NaC_2H_3O_2$ in 1 liter of 0.1 M $HC_2H_3O_2$.

$$H_3O^+ \text{ or } H^+, C_2H_3O_2^-, HC_2H_3O_2, Na^+, OH^-$$

Ans. $C_2H_3O_2^-$ most abundant
OH^- least abundant

11. Calculate the approximate H^+ ion concentration in 1 liter of 0.2 M acetic acid to which 0.2 mole of sodium acetate has been added (K_a for $HC_2H_3O_2 = 1.8 \times 10^{-5}$).

Ans. 1.8×10^{-5}

12. Explain how a buffer solution of NH_4Cl and NH_4OH would act toward the addition of a strong (a) acid and (b) base.

13. Calculate the OH^- ion concentration in 1 liter of 0.25 M NH_4OH which contains 0.2 moles NH_4Cl (K_b for $NH_4OH = 1.8 \times 10^{-5}$).

Ans. 2.25×10^{-5}

14. Calculate the approximate OH^- ion concentration in a solution which is made by adding 29 g of NH_4I to 1 liter of 2 *M* NH_4OH. (K_b for $NH_4OH = 1.8 \times 10^{-5}$.)

Ans. 1.8×10^{-4} g ion OH^-/liter

15. What weight of $NH_4C_2H_3O_2$ would be needed to maintain a H^+ ion concentration of 10^{-5} in 1 liter of 0.1 *M* $HC_2H_3O_2$ (K_a for $HC_2H_3O_2 = 1.8 \times 10^{-5}$)?

Ans. 0.18 mole or 13.86 g

16. A hypothetical weak base ionizes according to the equation

$$MOH \rightleftharpoons M^+ + OH^-$$

A salt (M_2SO_4) which is 100% ionized is added. The concentration of the base is 0.1 mole/liter, and the concentration of the M_2SO_4 is 0.05 mole/liter (K_b for the weak base $= 1 \times 10^{-6}$). Calculate (a) the OH^- ion concentration and (b) the H^+ ion concentration.

Ans. (a) 1×10^{-6} (b) 1×10^{-8}

17. The ionization of tetramethyl ammonium hydroxide proceeds according to the equation

$$(CH_3)_4NOH \rightleftharpoons (CH_3)_4N^+ + OH^-$$

($K_b = 4 \times 10^{-4}$.) For a 1×10^{-2} *M* solution, calculate (a) OH^- ion concentration and (b) H^+ ion concentration.

Ans. (a) 2×10^{-3}
(b) 5×10^{-12}

18. Given K_a for $HC_2H_3O_2 = 1.8 \times 10^{-5}$, (a) calculate K_H for potassium acetate, and (b) calculate the approximate OH^- ion concentration in a 0.1 *M* solution of $KC_2H_3O_2$.

Ans. (a) 5×10^{-9}
(b) 2×10^{-5}

19. NH_4CN is the salt of the weak acid NCN and the weak base NH_4OH (K_a for $HCN = 4.9 \times 10^{-10}$; K_b for $NH_4OH = 1.8 \times 10^{-5}$). Calculate K_H for NH_4CN.

Ans. 1.1

*p*H, *p*OH, and logarithms.

1. Perform the indicated operations by use of logarithms (see Appendix for four-place log tables).

(a) 418×653
(b) 8.62×14.3
(c) 96500×27.8
(d) 0.043×76.2
(e) $(4.3 \times 10^{-4}) \times (0.00035)$
(f) $\frac{6475}{35}$

(g) $\frac{4.48}{67.2}$

(h) $\frac{0.00083}{415}$

(i) $\frac{8.4 \times 10^{-2}}{2.1 \times 10^{-6}}$

(j) $\sqrt{7569}$

(k) $\sqrt[3]{636}$

(l) $\sqrt{12.96}$

(m) $(85.5)^3$

(n) $(34)^7$

(o) $(0.014)^5$

Ans. (a) 2.73×10^5
(b) 123
(c) 2.68×10^6
(d) 3.28
(e) 1.5×10^{-7}
(f) 185
(g) 6.7×10^{-2}
(h) 2×10^{-6}
(i) 4×10^4
(j) 87
(k) 8.6
(l) 3.6
(m) 6.25×10^5
(n) 5.25×10^{10}
(o) 6.8×10^{-10}

2. A solution contains 5×10^{-6} moles H^+ ion/liter.

   (a) Is the solution acid or basic?
   (b) What is the pH of the solution?

Ans. 5.3

3. (a) What is the *p*H of a solution in which the H^+ ion concentration is 4×10^{-4}?
   (b) What is the *p*OH of this solution?

Ans. (a) 3.4 (b) 10.6

4. (a) Calculate the concentration of H^+ and OH^- in 0.02 *M* NH_4OH ($K_b = 1.8 \times 10^{-5}$).
   (b) What is the *p*H and *p*OH of 0.01 *M* HBrO ($K_a = 2.1 \times 10^{-9}$).

Ans. (a) $[OH^-] = 6 \times 10^{-4}$
$[H^+] = 1.7 \times 10^{-11}$
(b) *p*H = 3.2, *p*OH = 10.8

5. Considering hydrolysis, indicate whether water solutions of the following salts would act acid, basic, or neutral.

   (a) NaCN
   (b) Na_3PO_4
   (c) NH_4Cl
   (d) $Mg(C_2H_3O_2)_2$
   (e) $ZnCl_2$
   (f) $Ca(NO_3)_2$
   (g) $CdCl_2$
   (h) $MnSO_4$
   (i) $CaBr_2$

6. The H^+ concentration of an acetic acid solution is 0.00030 *M*. What is the *p*H of the solution?

Ans. 3.5

7. Calculate the *p*H of 0.01 *M* HBrO ($K_a = 2.1 \times 10^{-9}$).

Ans. 5.8

8. A 0.02 *M* solution of a weak organic acid (HA) has a *p*H of 5.9. Calculate K_a for the weak acid.

Ans. 7.5×10^{-11}

9. What is the hydrogen ion concentration of a solution which has a *p*H of 9.77?

Ans. 1.7×10^{-10}

10. Assuming 100% ionization, *estimate* the *p*H of the following aqueous solutions.

(a) 0.1 *M* HCl
(b) 0.01 *M* NaOH
(c) 0.1 *M* Na_2CO_3
(d) 0.5 *M* NaOH
(e) 0.02 *M* HCl

Ans. (a) 1 (d) 13.7
(b) 12 (e) 1.7
(c) 8–9

11. Calculate the *p*H of 1.0 *M* benzoic acid (HBe) ($K_a = 6.4 \times 10^{-5}$).

Ans. 2.1

12. To buffer the solution in the previous exercise at a *p*H of 4.0, what weight of sodium benzoate (mol. wt. = 144) should be added to 1 liter of 1.0 *M* benzoic acid?

Ans. 92 g

13. What will be the *p*OH and *p*H of a buffer solution made up of 1.0 *M* N_2H_5OH and 0.50 *M* N_2H_5Cl [K_b (N_2H_5OH) $= 1.0 \times 10^{-6}$]?

Ans. *p*OH = 5.7, *p*H = 8.3

14. What is the approximate *p*H of a buffer solution made by mixing 1 liter of 0.3 *M* $HC_2H_3O_2$ and 1 liter of 0.6 *M* $NaC_2H_3O_2$ ($K_a = 1.8 \times 10^{-5}$)?

Ans. 5.05

15. Enough NH_4Cl is added to a 0.1 *M* solution of NH_4OH to make the solution 0.2 *M* with respect to NH_4^+. What is the OH^- ion concentration, *p*OH, and *p*H of the solution ($K_b = 1.8 \times 10^{-5}$).

Ans. *p*H = 8.95

16. K_a for the acid HA is 4.0×10^{-6}. Calculate the H^+ ion concentration and the *p*H of 1 *M* solution.

Ans. 2×10^{-3}; 2.7

17. A solution of NH_4OH was found to have an OH^- ion concentration of 2.0×10^{-4} g ions/liter. What is the *p*H of this solution?

Ans. 3.7

18. What are the *p*H's of solutions which contain:

(a) 0.046 g ion H^+/liter
(b) 0.0002 g ion H^+/liter
(c) 6.2×10^{-8} g ion H^+/liter

Ans. (a) 1.34 (b) 3.7 (c) 7.2

19. Indicate whether aqueous solutions of the following compounds are acidic, basic, or neutral.

(a) NH_4Cl
(b) $NaC_2H_3O_2$
(c) Na_2SO_4
(d) K_2CO_3
(e) $BaCl_2$
(f) NH_4NO_3
(g) $ZnSO_4$
(h) NaCN
(i) $Cu(NO_3)_2$
(j) K_3PO_4
(k) LiCl

20. Calculate the *p*OH and *p*H of the following solutions.

(a) 0.0063 g ion OH^-/liter
(b) 0.132 g ion OH^-/liter

Ans. (a) *p*OH = 2.2, *p*H = 11.8
(b) *p*OH = 0.9, *p*H = 13.1

21. $Ca(OH)_2$ is dissolved in H_2O to the extent of 0.0092 g/liter. It may be assumed to be 100% ionized. What is the (a) OH^- gram ion concentration, and (b) the *p*OH of the limewater solution?

Ans. (a) 2.5×10^{-4}
(b) 3.6

Solubility products.

1. Given K_{sp} for the following compounds,

Compound	K_{sp}	*Compound*	K_{sp}
(a) $BaCrO_4$	2.4×10^{-10}	(e) BaC_2O_4	2.2×10^{-7}
(b) FeS	3.7×10^{-19}	(f) $CaCO_3$	1×10^{-8}
(c) CdS	3.6×10^{-29}	(g) AgBr	8×10^{-13}
(d) HgS	4×10^{-53}	(h) NiS	1.4×10^{-24}

List the compounds according to increasing solubility, placing the least soluble compound first.

Ans. (d), (c), (h), (b), (g), (a), (f), (e)

2. Calculate K_{sp} of the following.

Salt	*Solubility in moles/liter*	Salt	*Solubility in moles/liter*
(a) $Ca(OH)_2$	0.021	(f) Ag_3AsO_4	2.4×10^{-5}
(b) AgCN	1.6×10^{-6}	(g) Ag_2SO_4	2.5×10^{-2}
(c) Ag_2CO_3	1.2×10^{-4}	(h) CaF_2	1.7×10^{-4}
(d) $AgIO_3$	3.1×10^{-4}	(i) $CaCO_3$	0.00013
(e) PbF_2	2.6×10^{-3}		

Ans. (a) 3.6×10^{-5}
(b) 2.6×10^{-12}
(c) 6.9×10^{-12}
(d) 9.6×10^{-8}
(e) 7×10^{-8}
(f) 9×10^{-18}
(g) 6.3×10^{-5}
(h) 2×10^{-11}
(i) 1.7×10^{-8}

3. From the solubilities of the following salts, calculate K_{sp}.

Salt	*Solubility in g/liter*	Salt	*Solubility in g/liter*
(a) $PbCO_3$	1.1×10^{-3}	(c) $CaSO_4$	1.93
(b) $SrSO_4$	0.114	(d) Ag_2CrO_4	3.2×10^{-2}

Ans. (a) 1.6×10^{-11}
(b) 3.8×10^{-7}
(c) 2×10^{-4}
(d) 3.5×10^{-12}

4. Calculate K_{sp} of the following.

Salt	*Solubility in mg/ml*
(a) AgCN	2.1×10^{-4}
(b) CaC_2O_4	6.8×10^{-3}
(c) As_2S_3	5.2×10^{-4}
(d) $PbCrO_4$	7×10^{-4}

Ans. (a) 2.6×10^{-12}
(b) 2.8×10^{-9}
(c) 4.4×10^{-27}
(d) 4×10^{-12}

5. The solubility of Ag_3AsO_4 is 8.5×10^{-3} g/liter. Calculate K_{sp}.

Ans. 2.8×10^{-18}

6. The solubility of the salt MX_2 is 3.0×10^{-4} moles/liter. Calculate K_{sp}.

Ans. 1.1×10^{-10}

7. The solubility of $Mg(OH)_2$ is approximately 0.009 g/liter. Calculate K_{sp}.

Ans. 1.5×10^{-11}

8. K_{sp} for $BaSO_4$ is 1.2×10^{-10}.

(a) What is the gram ion concentration of Ba^{++} per liter in a saturated solution?
(b) What is the number of grams of Ba^{++} per liter of saturated solution?
(c) What is the solubility of $BaSO_4$ in grams per liter?

Ans. (a) 1.1×10^{-5}
(b) 1.5×10^{-3}
(c) 2.6×10^{-3}

9. From the following table of K_{sp} values determine the approximate solubility of the compound in moles per liter.

Compound	K_{sp}
(a) $MgCO_3$	2.6×10^{-5}
(b) MnS	1.4×10^{-15}
(c) $Ni(OH)_2$	8×10^{-19}
(d) PbI	1.4×10^{-8}
(e) Ag_2SO_4	7.0×10^{-5}

Ans. (a) 0.005
(b) 3.7×10^{-8}
(c) 6×10^{-7}
(d) 1.5×10^{-3}
(e) 0.026

10. K_{sp} for Ag_2CO_3 is 6.1×10^{-12}. If a solution containing 4×10^{-5} g ions Ag^+/liter had 4×10^{-5} g ions CO_3^{--}/liter added to it, would precipitation occur? Show calculations to explain answer.

Ans. No

11. K_{sp} for $CaSO_4 = 2.3 \times 10^{-4}$. If 0.05 mole of Na_2SO_4 is added to 1 liter of solution containing 4×10^{-5} mole of calcium chloride, will $CaSO_4$ precipitate? Give a reason for your answer.

Ans. No

12. Calculate the minimum concentration of Cl^- necessary to precipitate AgCl from a solution in which the concentration of Ag^+ is 2.4×10^{-4} (K_{sp} for $AgCl = 1.2 \times 10^{-10}$).

Ans. 5×10^{-7}

13. K_{sp} for $Fe(OH)_3 = 1.0 \times 10^{-36}$. If a solution has a Fe^{3+} concentration of 1×10^{-3} M, what maximum concentration of OH^- can be present without precipitating $Fe(OH)_3$?

Ans. 1×10^{-11}

14. Na_2SO_4 was added to 0.01 M $BaCl_2$ until the concentration of SO_4^{--} was 0.0050 M. What concentration of Ba^{++} remained in solution? K_{sp} for $BaSO_4 = 1.2 \times 10^{-10}$.

Ans. 2.4×10^{-8}

15. An aqueous solution is 0.005 M $AgNO_3$. Solid NaCl is added until Na^+ in solution is 1.0 M. What is the concentration of Ag^+ left in solution? K_{sp} for $AgCl = 1.2 \times 10^{-10}$.

Ans. 1.2×10^{-10}

16. Equal volumes of 3.0×10^{-5} M $Ba(NO_3)_2$ and 2.0×10^{-5} M Na_2SO_4 are mixed. Will precipitation occur? Explain. K_{sp} for $BaSO_4 = 1.2 \times 10^{-10}$.

Ans. Yes, but very little

17. K_{sp} for $Ag_3PO_4 = 2.5 \times 10^{-18}$. Calculate the minimum concentration of PO_4^{---} necessary to start precipitation of Ag_3PO_4 from a 0.05 M solution of $AgNO_3$.

Ans. 2×10^{-14}

18. What would be the minimum gram ion concentration of S^{--} that would have to be present to start precipitation of Ag_2S from a 0.05 M solution of $AgNO_3$? K_{sp} for $Ag_2S = 1.6 \times 10^{-44}$.

Ans. 6.4×10^{-42}

19. A solution in equilibrium with a precipitate of $BaSO_4$ was found to contain 3.0×10^{-4} moles Ba^{++} per liter and 4.0×10^{-7} moles SO_4^{--} per liter of solution. Calculate K_{sp} for $BaSO_4$.

Ans. 1.2×10^{-10}

20. K_{sp} for CdS is 3.6×10^{-29}. A solution of $CdCl_2$ was found to contain 1.2×10^{-4} g ion Cd^{++}/liter. What concentration of S^{--} must be added to start precipitation?

Ans. 3×10^{-25}

21. A saturated solution of $SrSO_4$ contains 0.057 g $SrSO_4$/500 ml of solution. Calculate K_{sp} for $SrSO_4$.

Ans. 3.8×10^{-7}

22. Would $Al(OH)_3$ precipitate if the Al^{3+} ion concentration were 2.0×10^{-12} mole/liter and OH^- ion were added to make a OH^- ion concentration of 1.0×10^{-5}? Explain your answer with calculations ($K_{sp} = 4 \times 10^{-15}$).

Ans. No

23. K_{sp} for $Ag_2CrO_4 = 1.6 \times 10^{-12}$. If the CrO_4^{--} concentration in a solution is 1.3×10^{-4} moles/liter, what is the molar concentration of Ag^+ in contact with solid Ag_2CrO_4?

Ans. 1.1×10^{-4}

24. K_{sp} for $BaCrO_4 = 2.3 \times 10^{-10}$.

(a) What is the Ba^{++} gram ion content of a $BaCrO_4$ saturated solution?
(b) What is the actual weight in grams of Ba^{++} in a liter of saturated $BaCrO_4$ solution?

Ans. (a) 1.5×10^{-5} *M*
(b) 2.1×10^{-3} g/l

25. Calculate K_{sp} for $Th(SeO_4)_2$, given that 0.52 g dissolve per 100 ml.

Ans. 4×10^{-6}

26. The salt thallium (I) chloride, TlCl, is soluble in H_2O at 25° C to the extent of 3.60 g/liter. Calculate K_{sp}.

Ans. 2.3×10^{-4}

27. Given a 0.04 *M* solution of $AgNO_3$. K_{sp} of $Ag_3AsO_4 = 4 \times 10^{-18}$. What minimum gram ion concentration of AsO_4^{3-} would have to be attained to just start precipitation of Ag_3AsO_4?

Ans. 6×10^{-14} *M*

28. Calculate the approximate solubility (in moles per liter) of the compounds listed in the solutions indicated. Obtain K_{sp}'s needed from table in Appendix, page 218.

Compound	*Solution*
(a) AgBr	2×10^{-5} *M* NaBr
(b) $BaSO_4$	0.1 *M* Na_2SO_4
(c) AgCl	0.1 *M* HCl

Ans. (a) 1.7×10^{-8}
(b) 1.2×10^{-9}
(c) 1.2×10^{-9}

29. On saturation of 0.3 *M* HCl with H_2S, the S^{--} concentration is approximately 1.2×10^{-22} moles/liter. See page 218 for K_{sp} values.

(a) What concentration of each of the ions Cu^{++}, Cd^{++}, Mn^{++}, Fe^{++} will be needed to just start precipitation of the sulfide in each instance?
(b) Based on the values in (a), explain the separation of Cu^{++} and Cd^{++} in Group II from Mn^{++} and Fe^{++} in Group III in the cation analysis.

Ans. (a) 7×10^{-23} g ion Cu^{++}
1.3×10^{3} g ion Fe^{++}

30. (a) If the S^{--} ion concentration of a solution which is 0.3 *M* with HCl, and saturated with H_2S, is 1.2×10^{-22}, at what concentration of Pb^{++} will PbS just begin to precipitate? See Appendix, page 218, for K_{sp} values.
(b) At what concentration of Zn^{++} will ZnS just begin to precipitate?

Ans. (a) 3.5×10^{-6} (b) 0.1

31. Can a solution of $Cd(NO_3)_2$ in H_2O which will yield a precipitate with H_2S be made up (as in Problem 29)? Show calculations to verify your answer.

Ans. yes

32. Can a solution of $Mn(NO_3)_2$ in H_2O which will yield a precipitate with H_2S be made up (as in Problem 29)? Show calculations to verify your answer.

Ans. no

33. The *p*H of a saturated solution of $Mg(OH)_2$ is 10.6. Calculate K_{sp}.

Ans. 3.2×10^{-11}

34. K_{sp} for $Mg(OH)_2 = 3.2 \times 10^{-11}$ and K_{sp} for $BaCO_3 = 8.1 \times 10^{-9}$. How does the molar solubility of $Mg(OH)_2$ compare with that of $BaCO_3$?

Ans. $\frac{2 \times 10^{-4}}{9 \times 10^{-5}}$ or 1.8 to 1.0

Equations.

1. Balance the following equations from a consideration of electron transfer or change in oxidation numbers:

(a) $Ag_2S + HNO_3 \rightarrow Ag_2SO_4 + NO_2 + H_2O$
(b) $Ce_2(SO_4)_3 + H_2SO_4 + KMnO_4 \rightarrow Ce(SO_4)_2 + K_2SO_4 + MnSO_4 + H_2O$
(c) $HBr + NaMnO_4 + H_2SO_4 \rightarrow MnSO_4 + NaHSO_4 + Br_2 + H_2O$
(d) $H_2S + HBr + K_2Cr_2O_7 \rightarrow S + KBr + CrBr_3 + H_2O$
(e) $Mn(NO_3)_2 + HNO_3 + Pb_3O_4 \rightarrow Pb(NO_3)_2 + HMnO_4 + H_2O$
(f) $KI + KIO_3 + HC_2H_3O_2 \rightarrow KC_2H_3O_2 + I_2 + H_2O$
(g) $SbH_3 + HMnO_4 + H_2SO_4 \rightarrow MnSO_4 + H_3SbO_4 + H_2O$
(h) $V(SO_4)_2 + KMnO_4 + H_2O \rightarrow KHSO_4 + MnSO_4 + (VO_2)SO_4 + H_2SO_4$
(i) $Mo_2O_3 + KMnO_4 + H_2SO_4 \rightarrow MoO_3 + K_2SO_4 + MnSO_4 + H_2O$
(j) $SnCl_2 + H_2SO_3 + HCl \rightarrow SnCl_4 + H_2S + H_2O$

2. Balance the following equations, noting that more than two elements change in oxidation number.

(a) $As_2S_3 + HNO_3 \rightarrow As_2O_5 + NO + H_2SO_4 + H_2O$
(b) $Zn + Ag_3AsO_4 + H_2SO_4 \rightarrow ZnSO_4 + Ag + AsH_3 + H_2O$
(c) $MnI_2 + Pb_3O_4 + HNO_3 \rightarrow HMnO_4 + Pb(IO_3)_2 + Pb(NO_3)_2$
(d) $As_2S_3 + Mn(NO_3)_2 + Na_2CO_3 \rightarrow Na_3AsO_4 + Na_2MnO_4 + Na_2SO_4 + CO_2 + NO$
(e) $Sb_2S_3 + Na_2CO_3 + NaNO_3 \rightarrow Na_3SbO_4 + Na_2SO_4 + CO_2 + NO$

3. Balance the following ionic equations from a consideration of change in oxidation numbers or of electron transfer.

(a) $Sn^{++} + Fe^{3+} \rightarrow Sn^{4+} + Fe^{++}$
(b) $Cu + H^{+} + NO_3^{-} \rightarrow Cu^{++} + NO + H_2O$
(c) $H_2S + MnO_4^{-} + H^{+} \rightarrow SO_4^{--} + Mn^{++} + H_2O$
(d) $C_2O_4^{--} + Cr_2O_7^{--} + H^{+} \rightarrow Cr^{3+} + CO_2 + H_2O$
(e) $As_4 + H^{+} + NO_3^{-} \rightarrow AsO_3^{-} + NO + H_2O$
(f) $Bi(OH)_3 + SnO_2^{--} \rightarrow Bi + SnO_3^{--} + H_2O$
(g) $Fe + H^{+} + NO_3^{-} \rightarrow Fe^{3+} + NO_2 + H_2O$
(h) $Zn + H^{+} + NO_3^{-} \rightarrow Zn^{++} + NH_4^{+} + H_2O$
(i) $Ce^{4+} + AsO_3^{3-} + H^{+} \rightarrow AsO_3^{-} + Ce^{3+} + H_2O$
(j) $Pt + H^{+} + NO_3^{-} + Cl^{-} \rightarrow PtCl_6^{--} + NOCl + H_2O$

4. Balance the following by the ion-electron (half-reaction) method. Balance each step atomically and electrically; balance the gain and loss of electrons, then add the two steps to obtain the balanced single-step equation. Make all cancellations possible.

(a) $Sn^{++} \rightarrow Sn^{4+}$
$Ce^{4+} \rightarrow Ce^{3+}$

(b) $CdS \rightarrow Cd^{++} + S$
$H^{+} + NO_3^{-} \rightarrow NO + H_2O$

(c) $Bi_2S_3 \rightarrow Bi^{3+} + S$
$H^{+} + NO_3 \rightarrow NO + H_2O$

(d) $As_2S_3 + S^{--} \rightarrow AsS_4^{3-}$
$S_2^{--} \rightarrow S^{--}$

(e) $C_2O_4^{--} \rightarrow CO_2$
$H^{+} + MnO_4^{-} \rightarrow H_2O + Mn^{++}$

(f) $S_2O_3^{--} + H_2O \rightarrow SO_4^{--} + H^{+}$
$Br_2 \rightarrow Br^{-}$

5. Complete and balance the following equations by the ion-electron (half-reaction) method. Where necessary, add H^{+}, OH^{-}, or H_2O to complete and balance a half-reaction. Do not use H^{+} and OH^{-} together. Make all cancellations possible.

(a) $AsO_2^{-} \rightarrow AsO_4^{3-}$
$H^{+} + MnO_4^{-} \rightarrow H_2O + Mn^{++}$

(b) $Bi(OH)_3 \rightarrow Bi + OH^{-}$
$SnO_2^{--} \rightarrow SnO_3^{--}$

(c) $Cr^{3+} \rightarrow Cr_2O_7^{--} + H^{+}$
$ClO_3^{-} \rightarrow ClO_2$

(d) $Zn \rightarrow Zn(OH)_4^{--}$
$NO_3^{-} \rightarrow NH_3$

(e) $C_2H_5OH \rightarrow CH_3CHO$
$Cr_2O_7^{--} + H^{+} \rightarrow Cr^{3+}$

6. Balance the following equations (any method).

   (a) $V^{++} + MnO_4^- + H^+ \rightarrow Mn^{++} + VO_2^{++} + H_2O$
   (b) $NO_3^- + Al + OH^- + H_2O \rightarrow NH_3 + AlO_2^-$
   (c) $PbSO_4 + Na_2CO_3 + C \rightarrow Pb + Na_2SO_4 + CO_2$
   (d) $As_2S_5 + ClO^- \rightarrow AsO_4^{3-} + SO_4^{--} + Cl^-$
   (e) $Sb_2O_3 + Zn + H^+ \rightarrow SbH_3 + Zn^{++} + H_2O$
   (f) $CuCNS + MnO_4^- + H^+ \rightarrow Mn^{++} + Cn^- + Cu^{++} + S + H_2O$
   (g) $NpO_2^+ + H^+ \rightarrow Np^{4+} + NpO_2^{++} + H_2O$
   (h) $VO^+ + H^+ + Cr_2O_7^{--} \rightarrow VO_3^- + Cr^{3+} + H_2O$
   (i) $Sn^{++} + ClO_4^- + H^+ \rightarrow Sn^{4+} + Cl^- + H_2O$
   (j) $Pd + H^+ + Cl^- + NO_3^- \rightarrow PdCl_4^{--} + NOCl + H_2O$

7. Complete and balance.

   (a) $AgCl + NH_3 \rightarrow$
   (b) $Bi^{3+} + H_2S \rightarrow$
   (c) $K^+ + ClO_4^- \rightarrow$
   (d) $CO_3^{--} + H^+$ (excess) $\rightarrow$
   (e) $Ba^{++} + SO_3^{--} \rightarrow$
   (f) $Al^{3+} + OH^-$ (excess) $\rightarrow$
   (g) $PbS + HNO_3$ (dil.) $\rightarrow$
   (h) $Cd^{++} + NH_3 \rightarrow$
   (i) $SnS_3^{--} + H^+ \rightarrow$
   (j) $Sn^{4+} + Mg \rightarrow$

8. From the table of oxidation potentials on page 45 determine whether or not the following reactions will occur.

   (a) $Mn + Sn^{++} \rightarrow Mn^{++} + Sn$
   (b) $2\ Ag + 4\ H^+ + SO_4^{--} \rightarrow 2\ Ag^+ + H_2SO_3 + H_2O$
   (c) $2\ Na^+ + 2\ F^- \rightarrow F_2 + 2\ Na$
   (d) $6\ I^- + Cr_2O_7^{--} + 14\ H^+ \rightarrow 3\ I_2 + 2\ Cr^{3+} + 7\ H_2O$

Laboratory techniques.

1. What is meant by a "blank" test?

2. A precipitate is to be washed to reduce the amount of soluble reagent surrounding it. Consider the following possibilities:

   (a) one washing of a 3 ml portion of H_2O
   (b) three washings of 1 ml portions of H_2O

   Assume 0.5 mg of the reagent is present at the beginning of the washings. If this reagent is soluble to the extent of 0.5 mg/ml, and if after each washing 0.1 ml of liquid remains in the tube with the precipitate, calculate the milligrams of of the reagent left after each washing.

   Answers: (a) 0.017 mg; (b) after first washing 0.05 mg, after second washing 0.0045 mg, after third washing 0.00041 mg

3. In Problem 2, how many more times efficient is the triple washing with a total of 3 ml of wash liquid than was the single washing of 3 ml?

   Answer: about 42

Analysis for anions.

1. Explain clearly the necessity of making a prepared solution for anion analysis.

2. A prepared solution for anion analysis gave no precipitate with $BaCl_2 + NH_4OH$ or with $AgNO_3 + HNO_3$. Which of the following anions cannot be eliminated from specific tests?

   (a) PO_4^{3-} (b) Br^- (c) AsO_4^{3-} (d) NO_3^- (e) SO_4^{--}

3. An unknown solution containing only 1 anion gives a white precipitate with $BaCl_2$ and NH_4OH. The unknown solution also liberates a gas when treated with dil. HCl. When bubbled into $Ca(OH)_2$, the gas gives a white precipitate. A portion of the unknown when treated with $BaCl_2$ and HCl gives no precipitatate, but, when H_2O_2 is added to this solution, a white precipitate forms. What is the anion in the unknown?

4. In the chart below, indicate what happens to each of the anions when its sodium salt is

   (a) treated with conc. H_2SO_4
   (b) dissolved in H_2O, and NH_4OH and $BaCl_2$ are added
   (c) dissolved in H_2O, and HNO_3 and $AgNO_3$ are added

   (If no reactions occur, indicate.)

Anion	H_2SO_4 *Treatment*	$BaCl_2 + NH_4OH$	$AgNO_3 + HNO_3$
Br^-			
NO_3^-			
SO_4^{--}			
Cl^-			
PO_4^{3-}			
CO_3^{--}			
I^-			
$C_2H_3O_2^-$			
AsO_4^{3-}			
S^{--}			

Cation analysis.

Group I

1. In the precipitation of Group I with HCl, what would happen if the solution were very dilute?
2. If you were making up a solution of cations of the silver group, which salts would you use?
3. If the precipitate of Group I was insoluble in hot H_2O but completely soluble in NH_4OH, what conclusions could be drawn?
4. Using a single reagent, how could you distinguish between the pairs listed below? Indicate the reaction for each.

   (a) AgCl and $ZnCl_2$ (solids)
   (b) $PbCl_2$ and Hg_2Cl_2 (solids)
   (c) $Cd(NO_3)_2$ and $AgNO_3$ (solutions)
   (d) HCl and HNO_3 (solutions)
   (e) $Hg_2(NO_3)_2$ and $Hg(NO_3)_2$ (solutions)

Group II

1. In the precipitation of Group II, what would happen if the HCl were too concentrated? Too dilute?
2. Why is the precipitation carried out in 0.3 *M* HCl?
3. How is complete precipitation determined?
4. What conclusions may be drawn if the precipitate of Group II sulfides is completely soluble in Na_2S_2?
5. Why use dil HNO_3 rather than conc. HNO_3 in dissolving the sulfides of the copper group?
6. In precipitating $PbSO_4$ in Group II, why is it necessary to remove HNO_3 from the solution by evaporating to fumes of SO_3?
7. What is the basis for the separation of the following ions?

   (a) lead from copper, cadmium, and bismuth
   (b) bismuth from copper and cadmium
   (c) copper from cadmium

8. Give the formula for a reagent which will

   (a) dissolve CuS but not HgS.
   (b) dissolve $Cu(OH)_2$ but not $Bi(OH)_3$.
   (c) form a precipitate with Bi^{3+} but not with Cd^{++}.
   (d) dissolve $PbSO_4$.
   (e) form a precipitate with Cu^{++} but not with Cd^{++}.

9. What precautions should be taken in evaporating conc. H_2SO_4 to fumes of SO_3?

10. What conclusions may be drawn if

(a) only a white, slowly settling precipitate is produced on acidifying the Na_2S_2 solution?
(b) the precipitate of this group is completely soluble in conc. HCl?
(c) the precipitate of this group is yellow in color?

11. Give the formula for a reagent which will

(a) dissolve As_2S_5 but not HgS.
(b) dissolve Sb_2S_5 but not As_2S_5.
(c) form a precipitate with $HgCl_2$ but not with $SnCl_2$.

12. Diagram a scheme for the separation and confirmation of the following mixture of ions in H_2O. (Other cations may be assumed to be absent.)

(a) Hg_2^{++} (b) Cu^{++} (c) Sb^{3+}

Group III

1. What would happen if NaOH rather than NH_4OH were added in the precipitation of Group III with H_2S?

2. How should excess Na_2O_2 be disposed of?

3. What conclusions can be drawn if

(a) the precipitate of Group III dissolves completely in dil. HCl?
(b) no precipitate is obtained on addition of NaOH and H_2O_2 (or Na_2O_2)?
(c) the precipitate of Group III is white?
(d) the filtrate from the H_2O_2 (or Na_2O_2) treatment is colorless?
(e) the filtrate containing Group III is colorless?

4. If a test for iron is obtained in Procedure 10, how could you determine whether iron in the original sample was Fe^{++} or Fe^{3+}?

5. Why use HNO_3 rather than HCl in testing for manganese?

6. If the precipitate formed on addition of NaOH and H_2O_2 (or Na_2O_2) is brick-red, what conclusions might be drawn?

7. What difficulties would be encountered if

(a) chromium were incompletely oxidized by H_2O_2 (or Na_2O_2)?
(b) the solution to which $BaCl_2$ is added is too acid?
(c) the filtrate used in testing for Zn^{++} is too acid?

8. Give the formula for a reagent which will

(a) dissolve $Zn(OH)_2$ but not $Al(OH)_3$.
(b) dissolve ZnS but not S.
(c) form a precipitate with Al^{3+} but not Zn^{++}.

(d) form a precipitate with NaOH but not NH_4OH.
(e) form a precipitate with NH_4OH but not NaOH.
(f) dissolve $Ni(OH)_2$ but not $Fe(OH)_3$.
(g) dissolve $Al(OH)_3$ but not $Fe(OH)_3$.
(h) form a precipitate with Co^{++} but not Ca^{++}.

GROUPS IV–V

1. Explain the separation of Ba^{++} and Sr^{++} from Ca^{++}.
2. Why must a test for the ammonium ion be made on the original unknown?
3. Write the equation to show the action of a strong base on an ammonium salt.
4. Give the colors of the flame tests for Ca^{++}, Ba^{++}, Sr^{++}, K^+, and Na^+.
5. Explain how a cobalt glass assists in the flame test for K^+.
6. How would you differentiate the following three white salts.

   (a) $MgCl_2$ (b) NaCl (c) KCl
7. Why is it necessary to sublime ammonium salts before testing for K^+?
8. How would you distinguish between

   (a) NaCl and NH_4Cl (solids)
   (b) $MgCl_2$ and NH_4Cl (solids)
   (c) $MgCl_2$ and KCl (solids)
   (d) Mg^{++} and Na^+ (solutions)
9. Complete a diagrammatic scheme for the separation of and ultimate testing for a solution containing only Fe^{3+}, Al^{3+}, Sr^{++}, and Mg^{++}.

ALL GROUPS

1. From the following water-soluble salts—$CaCl_2$, K_3PO_4, $AgNO_3$, $CuSO_4$, $NaNO_3$—select three pairs which, when brought together in solution, will form a precipitate.

   (a) .... and ....
   (b) .... and ....
   (c) .... and ....
2. Give the formulas for two complex metallic ions of each of the following:

   (a) Ammonia (c) Sulfide (e) Cyanide
   (b) Hydroxide (d) Halide
3. Indicate a reagent with which you could distinguish between the following solids. Record reactions for each.

Solids	*Reagent*	*Results*
(a) Hg_2Cl_2 and $HgCl_2$	----------	----------
(b) CuS and HgS	----------	----------
(c) HgS and As_2S_5	----------	----------

4. Show by equations how a single reagent might be used to separate the following pairs of ions:

(a) Al^{3+} and Fe^{3+}
(b) Cu^{++} and Pb^{++}
(c) Hg^{++} and Zn^{++}
(d) Cr^{3+} and Fe^{3+}
(e) Ca^{++} and Mg^{++}
(f) Ag^{+} and Hg^{++}
(g) Hg_2^{++} and Hg^{++}
(h) Zn^{++} and Al^{3+}
(i) Ba^{++} and Sr^{++}
(j) Cd^{++} and As^{3+}
(k) Sn^{++} and Cu^{++}
(l) Bi^{3+} and Ni^{++}

Complex ions.

1. From the following list of cations, select those which form amphoteric hydroxides:

(a) Cd^{++}
(b) Cu^{++}
(c) Bi^{3+}
(d) Sn^{++}
(e) Al^{3+}
(f) Ni^{++}
(g) Cr^{3+}
(h) Zn^{++}
(i) Mg^{++}

Ans. (d), (e), (g), (h)

2. From the following elect those cations which form complex ions with NH_3:

(a) Cu^{++}
(b) Hg^{++}
(c) Cd^{++}
(d) Bi^{3+}
(e) Sn^{++}
(f) Cr^{3+}
(g) Ni^{++}
(h) Zn^{++}
(i) Al^{3+}
(j) Fe^{3+}
(k) Ag^{+}

Ans. (a), (c), (g), (h), (k)

3. From electronic configurations and orbital considerations, represent the formation of the following complex ions:

(a) $Cd(NH_3)_4^{++}$
(b) $AuCl_4^-$
(c) $FeCl_6^{3-}$
(d) $Pt(NH_3)_3Cl^-$
(e) $PdCl_6^{--}$
(f) $Co(CN)_6^{3-}$
(g) AsS_4^{3-}

5. Show by equations a reagent that will dissolve one but not both of the following pairs (all solids):

(a) $Al(OH)_3$ and $Fe(OH)_3$
(b) HgS and CuS
(c) $PbCl_2$ and AgCl
(d) $Zn(OH)_2$ and $Al(OH)_3$

(e) Na_2SO_4 and $PbSO_4$
(f) SnS and HgS
(g) As_2S_3 and CdS
(h) Sn and Cu
(i) Sb and Ni
(j) AgCl and AgI

6. Identify the following ions from the diagrammatic scheme shown below: A^+, B^{3+}, C^{++}, D^{3+}, and E^{3+}.

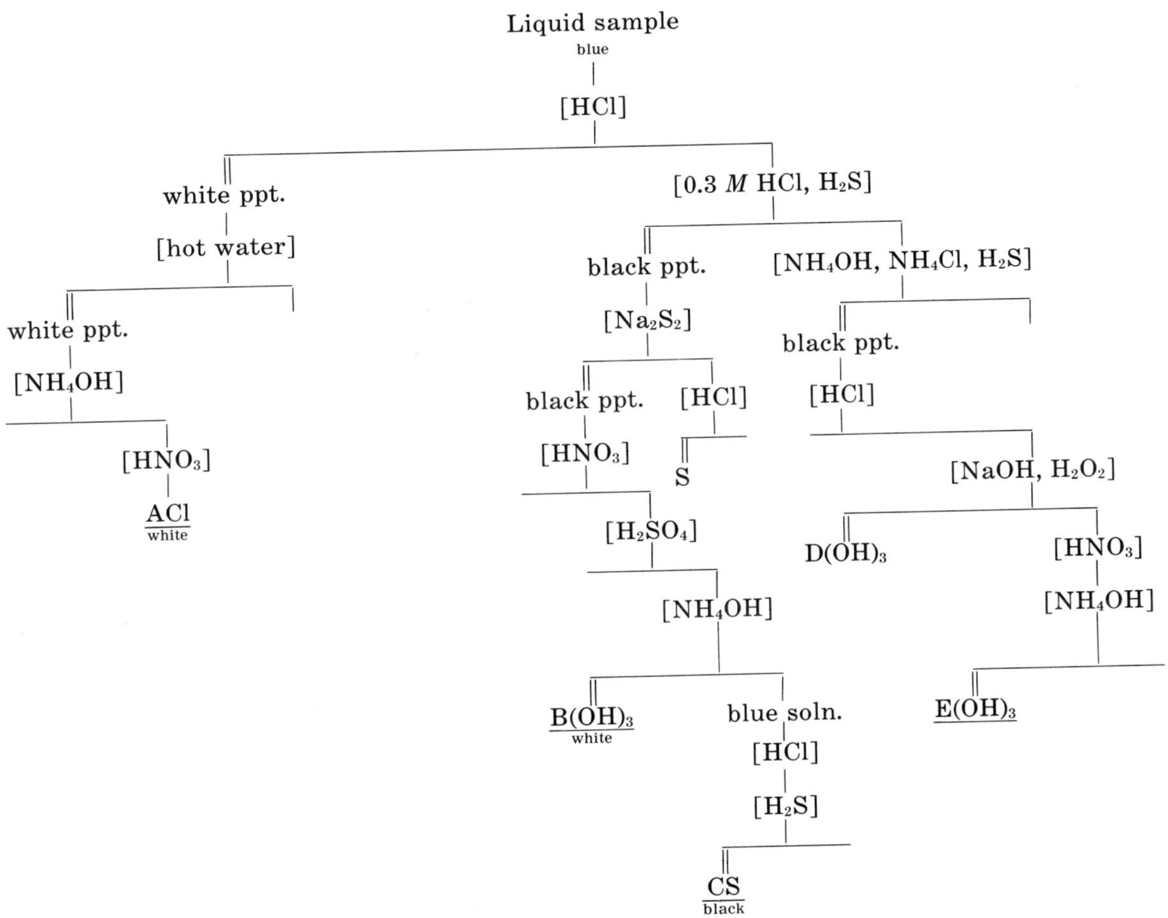

Alloy analysis.

1. Write equations for the reactions of conc. HNO_3 with the following metals.

   (a) Sn (b) Sb (c) Fe (d) Cu (e) Ag

2. How would you distinguish chemically between the following metals?

   (a) Cu and Sn
   (b) Pb and Bi
   (c) Fe and Ag
   (d) Cr and Cu
   (e) Mg and Ag

3. Show diagrammatically the separation and identification of metals in alloys of

   (a) Sn, Pb (b) Hg, Cu, Fe, Mg (c) Al, Cu, Sb

Analysis of salts and mixtures.

1. A sodium salt analyzed for anions gave the following results:

   (a) Solid + conc. H_2SO_4 gave a colorless gas with a sharp odor.
   (b) Aqueous solution gave no precipitate with $BaCl_2$ and NH_4OH, but gave a white precipitate with $AgNO_3$ and HNO_3.

   Therefore, the salt is Na.....................

2. A sodium salt analyzed for anions gave the following results:

   (a) Solid + conc. H_2SO_4—no visible change.
   (b) Aqueous solution + $BaCl_2$, NH_4OH—white precipitate soluble in HCl.
   (c) Solid + H_2SO_4 + methyl alcohol—combustible gas which burned with green flame.

   Therefore, the salt is Na.....................

3. The sulfide of a metal is insoluble in H_2O but soluble in cold dil. HCl. Its hydroxide is insoluble in H_2O but soluble in both NaOH and NH_4OH. What is the metal?

4. An unknown was found to be completely soluble in H_2O. Treatment of separate portions of the aqueous solution with Na_2HPO_4, $BaCl_2$, and $AgNO_3$ gave no precipitate. What cations and anions may be present?

5. An aqueous solution of a water-soluble white salt gives a precipitate with $(NH_4)_2CO_3$ in ammoniacal solution; it also gives a white precipitate with $AgNO_3$ and HNO_3. Which of the following salts is a possibility?

   (a) Hg_2SO_4 (b) NH_4Cl (c) $BaCl_2$ (d) $Al(NO_3)_3$ (e) $PbSO_4$

6. A salt is insoluble in water but dissolves in dil. HCl with effervescence of a colorless, odorless gas. A 0.3 *M* HCl solution of the salt gives a black precipitate with H_2S. Which of the following salts are possibilities?

   (a) $AlPO_4$ (b) NaBr (c) $KHCO_3$ (d) $SnCl_2$ (e) $PbCO_3$

7. A white solid is completely soluble in cold H_2O. Its aqueous solution is neutral to litmus. No change takes place when the solid is warmed with conc. H_2SO_4, and there is no change when the aqueous solution is tested with H_2SO_4 and $FeSO_4$. Which of the following are possibilities?

   (a) Na_2SO_4 (b) NiS (c) $PbCrO_4$ (d) $(NH_4)_2CO_3$ (e) KNO_3

8. An unknown mixture of salts is completely soluble in H_2O and contains CO_3^{--} and NO_3^- as the only anions. What cations may be present?

9. By means of a single reagent, how would you distinguish between the following pairs? Tell what happens in each case.

(a) CdS and HgS (solids)
(b) Sn^{++} and Hg^{++} (solutions)
(c) Hg_2^{++} and Hg^{++} (solutions)
(d) HCl and $HC_2H_3O_2$ (solutions)
(e) Bi^{3+} and Cu^{++} (solutions)

10. Tell how you would distinguish chemically between the following pairs of substances. In each case tell what happens to each substance.

Solutions	*Solids*
(a) $AgNO_3$ and $Zn(NO_3)_2$	(a) Hg_2Cl_2 and $HgCl_2$
(b) $Hg_2(NO_3)_2$ and $Hg(NO_3)_2$	(b) AgCl and $PbCl_2$
(c) $Pb(C_2H_3O_2)_2$ and $CuSO_4$	(c) Hg_2Cl_2 and $PbCl_2$
(d) $AgCl\cdot 2NH_3$ and $AgNO_3$	(d) AgCl and $BaCl_2$
(e) H_2SO_4 and HNO_3	(e) $BiCl_3$ and $ZnCl_2$
(f) HCl and HNO_3	(f) HgS and Bi_2S_3
(g) $HC_2H_3O_2$ and HCl	(g) PbS and SnS
(h) $HgCl_2$ and $SnCl_2$	(h) As_2S_5 and SnS_2
(i) $Pb(NO_3)_2$ and $CuSO_4$	(i) CuS and PbS
(j) $SnCl_2$ and $SnCl_4$	(j) CuS and Sb_2S_5
(k) $HgCl_2$ and $ZnCl_2$	(k) $Bi(OH)_3$ and $Pb(OH)_2$
(l) $(NH_4)_3AsS_4$ and $FeCl_3$	(l) FeS and NiS
(m) $CuSO_4$ and $Bi_2(SO_4)_3$	(m) FeS and MnO_2
(n) $Pb(NO_3)_2$ and $Bi(NO_3)_3$	(n) NiS and $Ni(OH)_2$
(o) $SbCl_3$ and $BiCl_3$	(o) $Zn(OH)_2$ and ZnS
(p) $PbCl_2$ and $BaCl_2$	(p) $AlCl_3$ and $Al(OH)_3$
(q) H_2S and HCl	(q) Na_2CrO_4 and $BaCrO_4$
(r) HCl and H_3AsO_4	(r) MnS and NiS
(s) $PbCl_2$ and $BiCl_3$	(s) NaCl and NH_4Cl
(t) $BiCl_3$ and $AlCl_3$	(t) $CaCO_3$ and Na_2SO_4
(u) $AlCl_3$ and $ZnCl_2$	(u) Ag and Zn
(v) $CrCl_3$ and $NiCl_2$	(v) Sn and Ni
(w) $ZnCl_2$ and $NiCl_2$	(w) Sn and Sb
(x) $MnCl_2$ and $ZnCl_2$	(x) Pb and Sb
(y) $FeCl_3$ and $FeCl_2$	(y) $BaCO_3$ and $BaSO_4$

11. A white salt (single) is completely soluble in cold H_2O and is also soluble in excess NaOH solution or excess NH_4OH. No precipitate is obtained with H_2S in 0.3 *M* HCl solution, but, when the solution is made basic with NH_4OH and H_2S added, a precipitate is obtained. The original sample treated with conc. H_2SO_4 yields a brown gas. An aqueous solution of the salt gives a pale yellow precipitate with $AgNO_3$ in HNO_3 solution. What is the salt?

12. An unknown mixture of salts is completely soluble in cold H_2O. The colorless solution gives a brick-red flame test. When the solution is treated with $AgNO_3$, no precipitate is obtained. What anions may be present?

13. Which of the following pairs of salts in aqueous solution would give no precipitate:

    (a) $BaCl_2 - AgNO_3$
    (b) $Hg_2SO_4 - Na_2S$
    (c) $NH_4Cl - CdSO_4$
    (d) $KBr - Pb(NO_3)_2$
    (e) $Na_3AsO_4 - Ca(C_2H_3O_2)_2$

14. A copper salt is insoluble in H_2O. A prepared solution (made by treating the solid with Na_2CO_3, filtering, and boiling the filtrate with dil. HNO_3) gave no precipitate with NH_4OH and $BaCl_2$. What anion(s) may be present?

31 Appendix

IONIZATION CONSTANTS OF WEAK ACIDS

Acid	K_i	*Apparent Percent of Ionization of 0.1 M Solution*
$H_2C_2O_4 \leftrightharpoons H^+ + HC_2O_4^-$	5.9×10^{-2}	77
$H_2SO_3 \leftrightharpoons H^+ + HSO_3^-$	1.7×10^{-2}	41
$HSO_4^- \leftrightharpoons H^+ + SO_4^{--}$	1.2×10^{-2}	35
$H_3PO_4 \leftrightharpoons H^+ + H_2PO_4^-$	7.5×10^{-3}	27
$HF \leftrightharpoons H^+ + F^-$	7.2×10^{-4}	8.5
$HNO_2 \leftrightharpoons H^+ + NO_2^-$	4.5×10^{-4}	6.7
$HC_2O_4^- \leftrightharpoons H^+ + C_2O_4^{--}$	6.4×10^{-5}	2.5
$HC_2H_3O_2 \leftrightharpoons H^+ + C_2H_3O_2^-$	1.8×10^{-5}	1.34
$H_2CO_3 \leftrightharpoons H^+ + HCO_3^-$	4.3×10^{-7}	0.21
$H_2S \leftrightharpoons H^+ + HS^-$	1.2×10^{-7}	0.11
$HSO_3^- \leftrightharpoons H^+ + SO_3^{--}$	1.0×10^{-7}	0.10
$H_2PO_4^- \leftrightharpoons H^+ + HPO_4^{--}$	6.2×10^{-8}	0.079
$HOCl \leftrightharpoons H^+ + OCl^-$	5.6×10^{-8}	0.075
$HCN \leftrightharpoons H^+ + CN^-$	4.0×10^{-10}	0.0063
$HCO_3^- \leftrightharpoons H^+ + CO_3^{--}$	4.7×10^{-11}	0.0022
$HPO_4^{--} \leftrightharpoons H^+ + PO_4^{3-}$	1×10^{-12}	0.0003
$HS^- \leftrightharpoons H^+ + S^{--}$	1×10^{-15}	0.00001

SOLUBILITY PRODUCTS

$AgBr$	3.5×10^{-13}
$AgCl$	1.2×10^{-10}
$AgCN$	2.2×10^{-12}
Ag_2CrO_4	1.7×10^{-12}
AgI	1.7×10^{-16}
$AgOH$	1.5×10^{-8}
Ag_3PO_4	1.8×10^{-18}
Ag_2S	1.6×10^{-44}
Ag_2SO_4	7.0×10^{-5}
$AgSCN$	1.1×10^{-12}
$Al(OH)_3$	4×10^{-15}
$BaCO_3$	4.9×10^{-9}
$BaSO_4$	1.2×10^{-10}
BaC_2O_4	1.7×10^{-7}
$BaCrO_4$	2.3×10^{-10}
BaF_2	1.7×10^{-6}
$CaCO_3$	1.7×10^{-8}
CaC_2O_4	3.8×10^{-9}
$CaCrO_4$	2.3×10^{-2}
CaF_2	3.2×10^{-11}
$CaSO_4$	2.3×10^{-4}
CdC_2O_4	1.5×10^{-8}
CdS	3.6×10^{-29}
$CuCl$	1.0×10^{-6}
CuS	8.5×10^{-45}
FeC_2O_4	2.1×10^{-7}
$Fe(OH)_2$	1.6×10^{-14}
$Fe(OH)_3$	1.1×10^{-36}
FeS	1.5×10^{-19}

SOLUBILITY PRODUCTS (Continued)

Hg_2Br_2	1.3×10^{-21}	$PbCl_2$	2.4×10^{-4}
Hg_2Cl_2	2.0×10^{-18}	$PbCO_3$	3.3×10^{-14}
Hg_2I_2	1.2×10^{-28}	$PbCrO_4$	1.8×10^{-14}
HgS	4×10^{-53}	PbI_2	1.4×10^{-8}
$MgCO_3$	2.6×10^{-5}	$Pb_3(PO_4)_2$	1.5×10^{-32}
MgC_2O_4	8.6×10^{-5}	PbS	4.2×10^{-28}
$MgNH_4PO_4$	2.5×10^{-13}	$PbSO_4$	2.3×10^{-8}
$Mg(OH)_2$	3.2×10^{-11}	$SrCO_3$	4.6×10^{-9}
MgF_2	6.4×10^{-9}	SrC_2O_4	1.4×10^{-7}
$Mn(OH)_2$	4.0×10^{-14}	$SrCrO_4$	3.8×10^{-4}
MnS	1.4×10^{-15}	SrF_2	2.8×10^{-9}
$Ni(OH)_2$	8.7×10^{-19}	$SrSO_4$	3.6×10^{-7}
NiS	1.4×10^{-24}	$Zn(OH)_2$	1.8×10^{-14}
		ZnS	1.2×10^{-23}

Equipment.

Individual equipment

1 beaker, 400 ml
1 beaker, 150 ml
2 beakers, 50 ml
1 test tube brush
1 box matches
1 Bunsen burner (micro) and tubing
1 burette clamp
1 test tube holder
1 crucible, No. 0
1 graduated cylinder, 10 ml
1 evaporating dish, 3″
1 file
1 pair forceps
5 pieces 6 mm glass tubing, 18″
6 pieces glass rod, 3 mm
1 spatula (metal)
1 iron ring, 4″
1 triangle
1 ring stand
1 test tube rack (for 3″ test tubes)
1 reagent block
14–15 ml dropping bottles
6 medicine droppers
18 3″ test tubes
1 watch glass
1 wire gauze
1 wing top
1 Florence flask, 250 ml
2 vials litmus paper
filter paper
3 towels
1 apron
1 platinum wire
1 sponge
1 pair safety glasses or goggles

Centrifuges, 1 for each 8–10 students

Set of reagents, liquid and solid (see below)

Sideshelf reagents for semimicro analysis.

(Liquids in 100 ml dropping bottles)

REAGENT	DIRECTIONS FOR MAKING SOLUTIONS
Acetic acid, 6*M*	350 ml glacial acetic acid and 650 ml H_2O
Alcohol, absolute, ethyl or Isopropyl alcohol, dry	100%

Reagent	Directions for Making Solutions
Aluminon	0.1% solution of ammonium aurin tricarboxylate
Ammonium acetate	250 g $NH_4C_2H_3O_2$ in 1 liter H_2O
Ammonium carbonate	250 g $(NH_4)_2CO_3$ in 1 liter 6 *M* NH_4OH
Ammonium chloride	100 g NH_4Cl in 1 liter H_2O
Ammonium hydroxide, 6 *M*	400 ml conc. NH_4OH and 600 ml H_2O
Ammonium hydroxide, 15 *M*	
Ammonium molybdate, 0.5 *M*	Dissolve 50 g MoO_3 in a mixture of 135 ml H_2O and 75 ml conc. NH_4OH. Add this solution slowly and with stirring to a mixture of 245 ml conc. HNO_3 and 575 ml H_2O. Makes 1 liter.
Ammonium oxalate	40 g $(NH_4)_2C_2O_4$ in 1 liter H_2O
Ammonium sulfate	132 g $(NH_4)_2SO_4$ in 1 liter H_2O
Arsenazo	0.001 *M* or 0.65 g/liter of $C_{16}H_{10}AsN_2Na_3O_{11}S_2 \cdot 2H_2O$
Barium chloride	120 g $BaCl_2 \cdot 2H_2O$ in 1 liter H_2O
Barium hydroxide	Saturated solution (calcium hydroxide may be used)
Carbon tetrachloride	
Chlorine water	Saturate water by bubbling chlorine gas through
Dimethyl glyoxime	1 g in 100 ml 95% ethyl alcohol
Dithizone test paper	Soak filter paper in a solution of 1 g dithizone in 25 ml chloroform and dry
Ether	
Ethyl alcohol	95%
Hydrochloric acid, 1 *M*	Dilute 80 ml 12 *M* HCl to 1 liter
Hydrochloric acid, 6 *M*	Dilute conc. HCl, 12 *M*, 1 to 1
Hydrochloric acid, 12 *M*	
Hydrogen peroxide	3% (Check for freshness or stability)
Iodine – potassium iodide mixture	Dissolve 168 g KI in 1 liter of H_2O and add 25 g iodine
Lanthanum nitrate	5% aqueous solution
Lead acetate	190 g $Pb(C_2H_3O_2)_2 \cdot 3H_2O$ in 1 liter H_2O
Magnesia mixture	Dissolve 130 g $Mg(NO_3)_2 \cdot 6H_2O$ and 240 g NH_4NO_3 in H_2O, add 35 ml 6 *M* NH_4OH, and dilute to 1 liter
Mercuric chloride	27 g $HgCl_2$ in 1 liter H_2O
Methyl alcohol	
Nitric acid, 16 *M*	
Nitric acid, 6 *M*	Dilute 375 ml conc. HNO_3 to 1 liter
α-Nitroso-β-naphthol	1 g in 100 ml 50% acetic acid
Potassium chromate	97 g K_2CrO_4 in 1 liter H_2O
Potassium cyanide	65 g KCN in 1 liter H_2O (KEEP IN STOREROOM)
Potassium ferricyanide	100 g $K_3Fe(CN)_6$ in 1 liter H_2O
Potassium ferrocyanide	105 g $K_4Fe(CN)_6$ in 1 liter H_2O
Silver nitrate	17 g $AgNO_3$ in 1 liter H_2O
Silver sulfate	5 g Ag_2SO_4 in 1 liter H_2O
Sodium carbonate	Saturated solution
Sodium acid phosphate	107 g Na_2HPO_4 in 1 liter H_2O
Sodium cobaltinitrite	Dissolve 25 g $NaNO_2$ in a mixture of 50 ml H_2O and 15 ml 6 *M* acetic acid. Add 2.5 g $Co(NO_3)_2 \cdot 6H_2O$. Allow to stand for 24 hours, then filter and dilute to 100 ml. (Fresh solution essential. Solid reagent may be stored.)
Sodium hydroxide, 5%	Dissolve 50 g NaOH in 1 liter H_2O
Sodium hypochlorite	Commercial bleach, e.g., "Purex"
Sodium polysulfide	Dissolve 480 g $Na_2S \cdot 9H_2O$ and 40 g NaOH in H_2O. Dilute to 1 liter. Dissolve 10 g flowers of sulfur in the solution.
Stannous chloride	Dissolve 100 g $SnCl_2 \cdot 2H_2O$ in 200 ml conc. HCl and dilute to 1 liter. Add a few pieces of metallic tin.
Sulfuric acid, 18 *M*	
Sulfuric acid, 3 *M*	Dilute 165 ml conc. H_2SO_4 to 1 liter
Thioacetamide	8% aqueous solution

Reagent	Directions for Making Solutions
Zinc uranyl acetate	Dissolve 10 g $UO_2(C_2H_3O_2)_2 \cdot 2H_2O$ in 6 ml 30% acetic acid with heat, if necessary, and dilute to 50 ml. Mix 30 g $Zn(C_2H_3O_2)_2 \cdot 3H_2O$ and 3 ml 30% acetic acid, and dilute to 50 ml. Mix solutions, add pinch of NaCl. After 12 hours, filter.

Cation test solutions. Test solutions of all cations should be available. The concentration of these test solutions should be approximately 10 mg of the ion per milliliter of solution (about 0.5 mg per drop).

Ion	Directions for Making Solutions
Ag^+	17 g $AgNO_3$ in 1 liter H_2O
Hg_2^{++}	Dissolve 14 g $Hg_2(NO_3)_2 \cdot 2H_2O$ in 100 ml 6 *M* HNO_3 and dilute to 1 liter
Pb^{++}	16 g $Pb(NO_3)_2$ in 1 liter H_2O
Cu^{++}	38 g $Cu(NO_3)_2 \cdot 3H_2O$ in 1 liter H_2O
Hg^{++}	17 g $Hg(NO_3)_2 \cdot \frac{1}{2}H_2O$ in 1 liter H_2O
Bi^{3+}	Dissolve 19 g $BiCl_3$ in 1 liter of 2 *M* HCl
Cd^{++}	28 g Cd $(NO_3)_2 \cdot 4H_2O$ in 1 liter H_2O
As^{3+}	Dissolve 14 g As_2O_3 in 1 liter 6 *M* HCl
Sn^{++}	19 g $SnCl_2 \cdot 2H_2O$ in 1 liter 6 *M* HCl
Sn^{4+}	30 g $SnCl_4 \cdot 5H_2O$ in 1 liter 6 *M* HCl
Sb^{3+}	19 g $SbCl_3$ in 1 liter 3 *M* HCl
Co^{++}	48 g $CoCl_2 \cdot 6H_2O$ in 1 liter H_2O
Ni^{++}	48 g $Ni(NO_3)_2 \cdot 6H_2O$ in 1 liter H_2O
Mn^{++}	54 g $Mn(NO_3)_2 \cdot 6H_2O$ in 1 liter H_2O
Zn^{++}	46 g $Zn(NO_3)_2 \cdot 6H_2O$ in 1 liter H_2O
Fe^{3+}	50 g $FeCl_3 \cdot 6H_2O$ in 10 ml conc. HCl and 1 liter H_2O
Al^{3+}	140 g $Al(NO_3)_3 \cdot 9H_2O$ in 1 liter H_2O
Cr^{3+}	51 g $CrCl_3 \cdot 6H_2O$ in 1 liter H_2O
Ba^{++}	20 g Ba $(NO_3)_2$ in 1 liter H_2O
Sr^{++}	33 g $Sr(NO_3)_2 \cdot 4H_2O$ in 1 liter H_2O
Ca^{++}	59 g $Ca(NO_3)_2 \cdot 4H_2O$ in 1 liter H_2O
Mg^{++}	105 g $Mg(NO_3)_2 \cdot 6H_2O$ in 1 liter H_2O
K^+	25 g KNO_3 in 1 liter H_2O
Na^+	25 g NaCl in 1 liter H_2O
NH_4^+	30 g NH_4Cl in 1 liter H_2O
Li^+	60 g LiCl in 1 liter H_2O

Unknown solutions. Unknown solutions should be made up to contain approximately 10 mg of each ion per milliliter of solution (see above).

Solids.

Ammonium chloride
Ammonium thiocyanate
Ferrous sulfate
Magnesium (granular or ribbon)
Methyl violet test paper
Oxalic acid
Potassium chlorate
Sodium acetate
Sodium arsenate
Sodium bismuthate
Sodium bromide
Sodium carbonate

Sodium chloride
Sodium chromate
Sodium iodide
Sodium nitrate
Sodium nitrite
Sodium peroxide
Sodium phosphate (Na_2HPO_4)
Sodium sulfate
Sodium sulfite
Sodium tetraborate
Zinc (granular)
Zinc sulfide or calcium sulfide

Four-place logarithm tables.

LOGARITHMS OF NUMBERS

Number	0	1	2	3	4	5	6	7	8	9
10	0000	0043	0086	0128	0170	0212	0253	0294	0334	0374
11	0414	0453	0492	0531	0569	0607	0645	0682	0719	0755
12	0792	0828	0864	0899	0934	0969	1004	1038	1072	1106
13	1139	1173	1206	1239	1271	1303	1335	1367	1399	1430
14	1461	1492	1523	1553	1584	1614	1644	1673	1703	1732
15	1761	1790	1818	1847	1875	1903	1931	1959	1987	2014
16	2041	2068	2095	2122	2148	2175	2201	2227	2253	2279
17	2304	2330	2355	2380	2405	2430	2455	2480	2504	2529
18	2553	2577	2601	2625	2648	2672	2695	2718	2742	2765
19	2788	2810	2833	2856	2878	2900	2923	2945	2967	2989
20	3010	3032	3054	3075	3096	3118	3139	3160	3181	3201
21	3222	3243	3263	3284	3304	3324	3345	3365	3385	3404
22	3424	3444	3464	3483	3502	3522	3541	3560	3579	3598
23	3617	3636	3655	3674	3692	3711	3729	3747	3766	3784
24	3802	3820	3838	3856	3874	3892	3909	3927	3945	3962
25	3979	3997	4014	4031	4048	4065	4082	4099	4116	4133
26	4150	4166	4183	4200	4216	4232	4249	4265	4281	4298
27	4314	4330	4346	4362	4378	4393	4409	4425	4440	4456
28	4472	4487	4502	4518	4533	4548	4564	4579	4594	4609
29	4624	4639	4654	4669	4683	4698	4713	4728	4742	4757
30	4771	4786	4800	4814	4829	4843	4857	4871	4886	4900
31	4914	4928	4942	4955	4969	4983	4997	5011	5024	5038
32	5051	5065	5079	5092	5105	5119	5132	5145	5159	5172
33	5185	5198	5211	5224	5237	5250	5263	5276	5289	5302
34	5315	5328	5340	5353	5366	5378	5391	5403	5416	5428
35	5441	5453	5465	5478	5490	5502	5514	5527	5539	5551
36	5563	5575	5587	5599	5611	5623	5635	5647	5658	5670
37	5682	5694	5705	5717	5729	5740	5752	5763	5775	5786
38	5798	5809	5821	5832	5843	5855	5866	5877	5888	5899
39	5911	5922	5933	5944	5955	5966	5977	5988	5999	6010
40	6021	6031	6042	6053	6064	6075	6085	6096	6107	6117
41	6128	6138	6149	6160	6170	6180	6191	6201	6212	6222
42	6232	6243	6253	6263	6274	6284	6294	6304	6314	6325
43	6335	6345	6355	6365	6375	6385	6395	6405	6415	6425
44	6435	6444	6454	6464	6474	6484	6493	6503	6513	6522
45	6532	6542	6551	6561	6571	6580	6590	6599	6609	6618
46	6628	6637	6646	6656	6665	6675	6684	6693	6702	6712
47	6721	6730	6739	6749	6758	6767	6776	6785	6794	6803
48	6812	6821	6830	6839	6848	6857	6866	6875	6884	6893
49	6902	6911	6920	6928	6937	6946	6955	6964	6972	6981
50	6990	6998	7007	7016	7024	7033	7042	7050	7059	7067
51	7076	7084	7093	7101	7110	7118	7126	7135	7143	7152
52	7160	7168	7177	7185	7193	7202	7210	7218	7226	7235
53	7243	7251	7259	7267	7275	7284	7292	7300	7308	7316
54	7324	7332	7340	7348	7356	7364	7372	7380	7388	7396

LOGARITHMS OF NUMBERS

Number	0	1	2	3	4	5	6	7	8	9
55	7404	7412	7419	7427	7435	7443	7451	7459	7466	7474
56	7482	7490	7497	7505	7513	7520	7528	7536	7543	7551
57	7559	7566	7574	7582	7589	7597	7604	7612	7619	7627
58	7634	7642	7649	7657	7664	7672	7679	7686	7694	7701
59	7709	7716	7723	7731	7738	7745	7752	7760	7767	7774
60	7782	7789	7796	7803	7810	7818	7825	7832	7839	7846
61	7853	7860	7868	7875	7882	7889	7896	7903	7910	7917
62	7924	7931	7938	7945	7952	7959	7966	7973	7980	7987
63	7993	8000	8007	8014	8021	8028	8035	8041	8048	8055
64	8062	8069	8075	8082	8089	8096	8102	8109	8116	8122
65	8129	8136	8142	8149	8156	8162	8169	8176	8182	8189
66	8195	8202	8209	8215	8222	8228	8235	8241	8248	8254
67	8261	8267	8274	8280	8287	8293	8299	8306	8312	8319
68	8325	8331	8338	8344	8351	8357	8363	8370	8376	8382
69	8388	8395	8401	8407	8414	8420	8426	8432	8439	8445
70	8451	8457	8463	8470	8476	8482	8488	8494	8500	8506
71	8513	8519	8525	8531	8537	8543	8549	8555	8561	8567
72	8573	8579	8585	8591	8597	8603	8609	8615	8621	8627
73	8633	8639	8645	8651	8657	8663	8669	8675	8681	8686
74	8692	8698	8704	8710	8716	8722	8727	8733	8739	8745
75	8751	8756	8762	8768	8774	8779	8785	8791	8797	8802
76	8808	8814	8820	8825	8831	8837	8842	8848	8854	8859
77	8865	8871	8876	8882	8887	8893	8899	8904	8910	8915
78	8921	8927	8932	8938	8943	8949	8954	8960	8965	8971
79	8976	8982	8987	8993	8998	9004	9009	9015	9020	9026
80	9031	9036	9042	9047	9053	9058	9063	9069	9074	9079
81	9085	9090	9096	9101	9106	9112	9117	9122	9128	9133
82	9138	9143	9149	9154	9159	9165	9170	9175	9180	9186
83	9191	9196	9201	9206	9212	9217	9222	9227	9232	9238
84	9243	9248	9253	9258	9263	9269	9274	9279	9284	9289
85	9294	9299	9304	9309	9315	9320	9325	9330	9335	9340
86	9345	9350	9355	9360	9365	9370	9375	9380	9385	9390
87	9395	9400	9405	9410	9415	9420	9425	9430	9435	9440
88	9445	9450	9455	9460	9465	9469	9474	9479	9484	9489
89	9494	9499	9504	9509	9513	9518	9523	9528	9533	9538
90	9542	9547	9552	9557	9562	9566	9571	9576	9581	9586
91	9590	9595	9600	9605	9609	9614	9619	9624	9628	9633
92	9638	9643	9647	9652	9657	9661	9666	9671	9675	9680
93	9685	9689	9694	9699	9703	9708	9713	9717	9722	9727
94	9731	9736	9741	9745	9750	9754	9759	9763	9768	9773
95	9777	9782	9786	9791	9795	9800	9805	9809	9814	9818
96	9823	9827	9832	9836	9841	9845	9850	9854	9859	9863
97	9868	9872	9877	9881	9886	9890	9894	9899	9903	9908
98	9912	9917	9921	9926	9930	9934	9939	9943	9948	9952
99	9956	9961	9965	9969	9974	9978	9983	9987	9991	9996